APPLIED MECHANICS Made Simple

The Made Simple series
has been created
primarily for self-education
but can equally well
be used as
an aid to group study.
However complex the subject,
the reader is taken,
step by step,
clearly and methodically
through the course. Each volume
has been prepared by
experts,
taking account of
modern educational requirements,
to ensure the most
effective way of
acquiring knowledge.

In the same series

Accounting
Acting and Stagecraft
Additional Mathematics
Administration in Business
Advertising
Anthropology
Applied Economics
Applied Mathematics
Applied Mechanics
Art Appreciation
Art of Speaking
Art of Writing
Biology
Book-keeping
British Constitution
Business and Administrative Organisation
Business Economics
Business Statistics and Accounting
Calculus
Chemistry
Childcare
Commerce
Company Administration
Company Law
Computer Programming
Computers and Microprocessors
Cookery
Cost and Management Accounting
Data Processing
Dressmaking
Economic History
Economic and Social Geography
Economics
Effective Communication
Electricity
Electronic Computers
Electronics
English
English Literature
Export
Financial Management
French
Geology
German
Housing and Planning Law
Human Anatomy
Italian
Journalism
Latin
Law
Management
Marketing
Mathematics
Modern Biology
Modern Electronics
Modern European History
New Mathematics
Office Practice
Organic Chemistry
Philosophy
Photography
Physical Geography
Physics
Pottery
Psychiatry
Psychology
Rapid Reading
Retailing
Russian
Salesmanship
Secretarial Practice
Social Services
Soft Furnishing
Spanish
Statistics
Transport and Distribution
Twentieth-Century British History
Typing
Woodwork

APPLIED MECHANICS Made Simple

George E. Drabble, BSc(Eng)

Made Simple Books
W. H. ALLEN London
A Howard & Wyndham Company

Made and printed in Great Britain
by Richard Clay (The Chaucer Press), Ltd., Bungay, Suffolk
for the publishers W. H. Allen & Co. Ltd.,
44 Hill Street, London, W1X 8LB

First edition, June 1971
Reprinted (with additional exercises), September 1977
Reprinted, August 1980

SBN 0 491 00208 4 paperbound

Foreword

Applied Mechanics Made Simple is intended to present the fundamental principles of Mechanics in a manner that can be easily understood, and to show how they are applied to engineering problems. The first part of the book discusses these principles themselves: the principles of Statics, known for thousands of years, and the principles of Dynamics, embodied in the famous Laws of Newton. The remainder of the book discusses particular aspects of these ideas. Statics is investigated in terms of friction, the design of structures, stress and strain, and fluid pressure, whilst dynamics is studied in the context of rigid bodies and also of fluids.

The book is not intended as a textbook, but should form useful background reading for students taking the first year of an Engineering Degree course, and also for school-leavers who wish to know more of this subject before committing themselves to a course of study. It is particularly appropriate for H.N.C. and similar courses, in giving an excellent insight into this essential aspect of engineering. Every attempt has been made to keep explanations at the simplest possible level, and a minimum of mathematical analysis has been used. Examples have been included, and these have all been calculated in the MKS system of units.

The author owes a particular debt of gratitude to Professor D. S. Dugdale and Mr. P. Polak of the University of Sheffield for reading and criticising the original manuscript, for reading the proofs and for their help and encouragement throughout the production of the book.

G. E. DRABBLE

Table of Contents

CHAPTER ONE

INTRODUCTION

(1) What is Applied Mechanics?

Mechanics is the study of objects, or bodies, as we shall call them, when subjected to forces. *Applied* mechanics is the application of the theories of mechanics to real engineering situations. A knowledge of applied mechanics is thus essential to students and practitioners of engineering.

In studying the effects of forces on bodies, problems fall into two broad general classifications. We have, first, all those cases in which the motion of the bodies is not affected by the forces we are studying. The simplest illustration of this is a structure such as a bridge. A bridge is certainly subjected to forces; it has to withstand a certain load, for a start. Furthermore, it has its own weight to support, and in a large bridge this may be many times the value of the load it is to carry. But these forces do not affect the motion of the bridge, which is stationary at all times. The study of forces in such a situation is called **statics**. Another example would be an aircraft in straight level flight at constant speed. The aircraft is subjected to the thrust of the engine, its own weight, the lift due to the air, and the frictional drag of the air. But these forces, when taken all together, do not affect the motion of the craft, which is steady and constant. So statics is not necessarily confined to the study of stationary bodies.

Secondly, we have all those situations in which forces contribute towards a change of motion of the body being investigated. An aircraft at take-off or landing, a rocket being launched, a piston moving in a cylinder, or the Moon moving round the Earth, are examples of motion which is not constant. The study of forces under such conditions is called **dynamics**. In order fully to understand what is happening in such situations, it is necessary to be quite clear about our ideas of motion itself, without relation to the forces. For example, many readers might not appreciate that the motion of the Moon around the Earth involves a continuous change of velocity. So dynamics, in turn, is sub-divided into **kinematics**, which is the study of motion itself, and **kinetics**, the study of motion in relation to the forces causing it.

Applied mechanics, then, is the study of the relationship between forces and all fixed and moving objects. This is a pretty wide field, varying from the movements of stars and galaxies down to the movements of atoms and electrons. In a book of this kind, we can hardly be expected to cover so wide a range; to do so would require at least an elementary knowledge of Relativity and of Quantum Mechanics. But if we are prepared to exclude those bodies which travel at speeds close to the speed of light, and those bodies of submicroscopic size, we shall find that the principles governing the effects of forces on the vast remainder of moving bodies are extremely few in number and simple to understand.

(2) The Plan of this Book

We start by examining motion itself, and the quantities that are required to define the motion of various bodies. We shall see in Chapter Two, when we study kinematics, that displacement, velocity and acceleration, when accurately defined, are sufficient to fully determine the motion of all bodies that come within the range of ordinary practical engineering.

In Chapter Three we study forces themselves, and the laws governing the combining of two or more forces. At this stage we are not concerned with the effects of forces but only with how we can manipulate them. This is the province of statics. Statics may perhaps claim to be one of the oldest of the applied sciences. It is not, according to our definition, a branch of applied mechanics at all, but it forms a necessary preliminary to the study of it, much as the moves of the various pieces must be learned before you can play chess.

In Chapter Four we come to kinetics. Force and motion are linked together by the three laws of motion formulated by Isaac Newton. Newton's Laws are just as valid today as they ever were for bodies moving at moderate speeds—which we may assume takes us up to many thousands of miles per hour. Newton's Laws were not exploded by Einstein—they were verified. Einstein extended the range of dynamics, to include bodies moving at speeds which had not been contemplated in Newton's time, and for which the famous Laws were inaccurate. But for all other bodies, Einstein's and Newton's Laws are for all practical purposes identical. Chapter Five deals with the concepts of work, energy, power and momentum. All these quantities may be thought of as stemming from Newton's Laws, and will be found useful in dealing with particular aspects of force and motion.

Chapters Six and Seven deal with two particular branches of statics. Friction, that force which is present in all moving machinery, is examined in detail in Chapter Six, while Chapter Seven deals with the structural engineer's job of examining forces, not on, but within a structure.

Chapters Eight, Nine and Ten could be grouped together under the general title of Applied Statics. We are concerned here, not with the effect of forces on the motion of bodies, but the effect on the bodies themselves: how bodies withstand force, and how they deform under force. We examine the concepts of stress and strain, and analyse two important applications of these concepts: the bending of beams and the twisting of shafts.

Finally, we take a brief look at what might superficially appear to be a completely different field: the study of forces applied to fluids. We shall see that the principles of statics can be effectively used to determine forces set up by the pressure of water on dam walls and similar situations, and also that the principles of energy and momentum are valid for dealing with the flow of fluid in pipes and jets.

(3) Applied Mechanics as a Science

There is a tendency among non-scientific people to regard an applied science as no more than the derivation of a number of formulae, and the application of these to physical situations. This is certainly one aspect of applied mechanics, and it should not be underestimated. But neither should it be invested with too much importance. A formula is probably the best way of conveying a great deal of technical information. But unless the principles governing the derivation of the formula are fully understood by those whose job it is to use it, the situation may arise where the formula is being used in circumstances where it is not valid. To learn about applied mechanics is not only to learn a number of formulae; it is also to understand how they are derived, to understand the laws on which they are based, and to understand how these laws came to be stated in the first place. Applied mechanics is concerned with real objects: aeroplanes, ships, bridges, spacecraft and so on, and one should never lose sight of the fact that the laws and formulae used in its practice are themselves based on experiments on real objects.

Newton's Laws in dynamics, Hooke's Law in stress analysis, and the Law of Conservation of Energy are not merely abstract ideas that have been thought up as academic exercises in the manner of some of the ancient Greek philosophies. They are every one founded on a basis of sound experimental observation. Einstein's principles of relativity were not merely just thought up, but were an attempt to explain certain otherwise unaccountable errors in an experiment to determine the velocity of light. What is more, these principles were not universally accepted for many years—until 1919, in fact, when certain predictions based on them were verified by direct observation. The essence of science is experiment. In mathematics, it is quite possible to build up a logical system based on pure imagination, and to develop arguments, theorems and conclusions which are based entirely on the original axioms. Euclidean geometry is a classical example. But this is not science in the modern sense. Science nowadays is concerned with the investigation of physical facts and the formulation of general theories arising out of these investigations.

The chain of development starts with a hypothesis. This is a guess on someone's part—a hunch, if you like—that something is related to something else in some way. Suppose, for instance, I put forward the hypothesis that large bodies fall to earth faster than small ones. There is something to be said for such a hypothesis. We know that a brick falls faster than a feather. To test the hypothesis is (or should be) the next step. I should collect a large number of objects and arrange them in order of size. Now comes the first difficulty. What do I mean by size? I can assert that size is, for my purpose, the same as volume. So I arrange my objects in order of volume, and then drop them all from the same height. It is not long before I have disproved the hypothesis. I find that an apple falls much faster than a balloon.

So I try a second hypothesis. I suggest that heavy bodies fall faster than light ones. I now perform my second experiment, rearranging my objects in order of weight, and dropping them again. This time I have much more success, although I find I have to take greater care over my measurements. I find, for instance, that the time of fall of a brick is only a fraction more than that of a cotton-reel. In fact, as I increase the accuracy of my time measurement, sooner or later I encounter conflicting evidence. A polished steel ball has only one-tenth the weight of a brick, but falls faster. So my second hypothesis has to be abandoned.

In arriving at a third hypothesis, I begin to guess that the presence of air is a relevant factor. Eventually it may come to me, either as a result of long and careful thought or as a flash of inspiration, that if I remove the air, all the bodies would probably fall at the same rate. This is the third hypothesis. I must now take a collection of objects, such as marbles, grains of corn and feathers, place them in a tube, evacuate the air, and cause them to fall. If they reach the other end of the tube together, I am now able to replace the hypothesis by a theory. There are two good reasons for this. First, the final experiment shows no deviations. Secondly, the deviations in the earlier experiments can now all be explained satisfactorily by the presence of the air. The final theory can now be stated: 'All bodies, of whatever size or weight, would fall to the earth at a given point with the same acceleration, in the absence of the air'. This would suffice as a theory until it was shown to have exceptions, or until it was seen to be part of a broader generalization.

If this sounds obvious and simple, think how many false theories have stood the test of time just because people would not bother to test them. Aristotle is sometimes revered as a great philosopher, but he maintained, along with a great deal of other fiction, that women had fewer teeth than men. (And Aristotle was twice married.) I wonder how many readers are convinced that the summers were much warmer, and the winters much colder, when they were children; and how many have bothered to hunt up the figures to find out.

All this is harmless and amusing, but if your business is to establish theories for general application, so that others may determine the sizes of the various components of an air-liner, or a large dam, it is necessary to put personal whims aside and to be sure that your theories are soundly based and amply proved by experiment.

(4) Approximations and Assumptions

In formulating my theory of falling bodies in the previous section, I assumed no air to be present. But if this theory is to have practical application, we cannot just get rid of the air. Whoever makes use of it must do so with discretion: he must be aware of the presence of air, and must realize that falling bodies will not behave exactly according to the theory. In formulating laws and in deriving theories it is always necessary to assume certain ideal conditions which may not obtain in practice. It is not easy to judge the extent to which the ideal theory may be expected to predict the non-ideal situation.

Let us imagine two superficially similar cases: the dropping of a bomb from an aircraft and the firing of a cannon shell. Both are problems of motion which can be solved by the equations derived from Newton's Laws. Unfortunately, air is present in both cases, and we have no accurate way of estimating the effect of the air. We will suppose that we wish to know the time of fall of the bomb from the aircraft, and also the range of the shell. Assuming no air, the appropriate equations give us 5 seconds and 800 metres, respectively. The persons most interested in the answer to the bomb problem would be the crew of the aircraft, who very reasonably want to be well away from the scene when the explosion occurs. We judge that the effect of the air would be to increase the time of fall of the bomb, but only by a small amount. In any case, we think that 5 seconds is ample time—the plane is probably half a mile away in that time—and furthermore, the error of our calculation is an error on the safe side.

However, with the shell problem we realize that, because of the presence of air, our figure of 800 metres is too large. The actual distance might be substantially less than 800 metres, and the error here is on the wrong side. In such a case, we could either perform a much more careful and rigorous analysis, which took into account the effect of the air, or (as is more likely) determine the range of the shell by some other means—possibly by experiment.

Every theory and every formula in applied mechanics relies on certain assumptions and certain simplifications. The user has to be aware of these, and must judge whether they are significant enough to affect his calculations seriously, and if so, by how much and in which direction. We may ignore the resistance of the air if we are investigating the flow of a river, but not if we are predicting the flight of a shell. For the purpose of determining the flow of oil in a pipeline, we may assume the oil to be incompressible; but we plainly cannot assume that is is incompressible if it is to be used in the cylinder of a shock-absorber. In studying the motion of a lift or hoist, we may frequently assume that the weight of the supporting cable is negligible. But the weight of the cable of a mine hoist might well be many times that of the cage itself, and to ignore it would be absurd.

One might ask why we should neglect these things if they are there, and if sometimes we have to take them into account anyway. The answer is that, by making such assumptions, the mathematical analysis is very much simplified. We forfeit accuracy for the sake of simplicity. Many readers must have encountered the simple pendulum at school. In analysing the motion of the pendulum, at least three assumptions are made: the mass is assumed all concentrated at the centre of gravity; the resistance of the air is neglected; the sine of the angle of swing is approximated to the angle in radian measure. As a result, we obtain a very simple formula for the period of swing of the pendulum *which is probably accurate to 99·9 per cent.* A more exact analysis would be possible, but would be extremely complex. For all practical applications, the gain in accuracy would not justify the increased complexity.

I shall refer to this matter constantly throughout this book. Mathematics and theoretical mechanics is concerned with model systems operating under perfect conditions. Applied mechanics is concerned with obtaining realistic and useful answers in real situations. In calculating the stress in a member of a structure, a designer is not worried if, in practice, the actual value is 95 per cent, or even 105 per cent, of his predicted figure—he probably allows a safety factor of 3 anyway. An automobile engineer who calculates that a car requires an engine capable of a tractive force of 600 pounds may arrive at this figure on an assumed speed of 60 miles per hour. If the actual required force at this speed is 575 pounds, the car will merely travel a little faster, that is all.

(5) Units and Dimensions

The measurement of physical quantities is the first step in formulating the laws of applied mechanics. It is true that certain crude empirical laws can be formulated without the necessity of performing measurements, but these laws are of little practical use to the designer or the research worker. For example, one may state a law such as 'All heavy bodies drop downwards'. This statement is useful to a limited extent in that our knowledge of the fact helps us in certain situations; for instance, when crawling under a car, we take good care that it is well supported on the jacks. But it can hardly qualify for the dignified status of a law. Aside from other considerations, it begs the question as to what we mean by 'heavy'. The obvious answer is that a heavy body is one which drops downwards. Furthermore, the idea of 'downwards' is of little use to an astronaut. As a contrast to this, the famous Boyle's Law of gases states: 'If the temperature of a gas is kept constant, the pressure and the volume will be inversely proportional.' But implicit in this law is an accurate and agreed method of measuring temperature, pressure and volume.

The study of applied mechanics involves the use of scores of measured quantities. Length, speed, force, pressure, volume, time, acceleration and stress are just a few of these. Some of them are fundamental in that they require a set of units, and a measuring device based on an arbitrary standard. For example, we may choose an arbitrary unit of length of one metre (originally a fraction of the length of the circumference of a quadrant of the Earth), and we can then make scales, rules and surveyor's chains in this unit, together with its multiples and fractions. Similarly, our unit of time can be taken as the second (originally a fraction of the length of the sidereal day), and stop-watches and other instruments can be made to measure time by this standard. But some measurements may not need any further measuring device. A simple instance is that of speed. The speed of a car can be determined by a combination of a measurement of length and one of time—indeed, this is actually done when vehicles under test are timed over a measured length with a stop-watch. The calculated speed may then be expressed in terms of the units of its components (for example, so many metres per second) or it may be given a different name, for the sake of brevity. (In motoring circles, a speed of a hundred miles an hour is sometimes, rather obscurely, called a ton.) A quantity which is determined entirely by the measurement of other quantities is called a **derived** quantity, as distinct from the **fundamental** quantities it is based upon.

It is important to realize that the choice of fundamental quantities is quite arbitrary. In the example I have given, a derived quantity of speed was based on the two fundamental quantities of length and time. But it would be just as reasonable, and in some respects, more so, to choose speed as a fundamental quantity, basing the unit on, say, a fraction of the speed of light. The quantity of length would then become a derived quantity: the basic unit of length would be the distance covered at unit speed in unit time (i.e. one second).

Rather surprisingly, it can be shown that all the measurements required in applied mechanics can be expressed in variations of only three fundamental quantities. These are usually chosen arbitrarily as **length**, **time**, and **mass**, although some writers adopt force as an alternative to mass. At the time of writing, efforts are being made to establish an internationally accepted rational system in which the units of the three fundamental quantities are:

Length:	**metre**
Mass:	**kilogramme**
Time:	**second**

Although these units may be tolerably familiar to most readers, many of the corresponding derived units will perhaps not be so. To speak of the speed of a car as 30 metres per second may convey only a vague idea of the magnitude of this speed to people accustomed to reading speed in miles per hour on a speedometer. For this reason, in making illustrative calculations I shall use the values based on the three fundamental units chosen, but frequently I shall also supply the more familiar and popular alternative.

A special word is necessary about the unit of force. It will be seen in Chapter Four that force is actually a derived quantity, based on mass, length and time. Weight, which is the gravitational attraction between two bodies, is a special case of force, the force in such a case being directly proportional to the mass of the body. Now force is relatively easy to measure—a simple spring balance will measure force with high accuracy—but mass is not so easy. Indeed, if you read Chapter Four attentively, you may be able to deduce that, in the absence of a gravitational field, the only method of determining mass is to observe the motion of the said mass under known conditions of applied force. One convenient short cut for earth-bound individuals, however, is to weigh the mass. As long as the limitation of this method is appreciated, it is a very good method (indeed the best one) for evaluating mass.

If you weigh the body using a spring balance, you are measuring the gravitational pull on it, which is directly proportional to the mass. If you weigh it using a beam balance, you are comparing the weight of your mass with another mass of known value. This second method of finding mass would be equally effective on the Moon as on the Earth; the trouble is that it has inevitably caused some confusion between the quantities of mass and of weight. In most cases, this has not mattered. To speak of the weight of a kilogramme mass as 'one kilogramme', or the weight of a load of coal as 'one ton', or the weight of a bag of flour as 'seven pounds', seems only logical and sensible to most people.

Unfortunately, because we have already chosen to accept units of length, mass and time, we have to accept the unit of force implicit in these chosen units, and this fact has always caused trouble to the student of applied mechanics. At the outset, he has to arrive at an entirely new concept of force and mass in which the weight of a given mass has a name and number which differs from the name and number of the mass. He has to resist constantly the temptation to express the weight of a body in terms of its mass.

Most people nowadays find it easy to accept that the weight of a body is not an absolute quantity in the same way that its mass is, but varies according to its position in space. The simple rule (which I shall explain later on) is that the weight of a body is determined by multiplying the mass by a quantity called the **gravitational acceleration**, usually denoted by g. This quantity depends on the location of the body: whether it is on the Moon, or the Earth, or even what part of the Earth it is on. In the system of units we shall use throughout this book, the gravitational acceleration for masses anywhere on the Earth is approximated to 9·81 metres per second per second, and it means that the weight of a kilogramme is 9·81 units of force. This unit of force has been allocated the name of the **newton**. For practical purposes, a newton may be visualized as a little less than a quarter of a pound weight.

For most people, this esoteric idea of weight is unnecessary and unimportant. Those whose interests and activities lie outside the realms of applied mechanics, such as grocers, politicians, artists and farmers, may choose to express weight in units which we shall reserve for mass; they are doing nobody any harm, and are saving themselves and their associates a great deal of bother and trouble. But if you want to become acquainted with the rudiments of applied mechanics, this is one of the things you have to learn to come to grips with, and try to understand.

(6) Mathematics and Special Terms

Although an understanding of applied mechanics calls for a certain amount of mathematics, the fundamental principles require only a very limited knowledge. All the work in this book can be understood by a reader with a knowledge of the basic processes of arithmetic, an elementary knowledge of algebra (sufficient to understand the significance of a quadratic equation, even though insufficient to solve it) and a very elementary knowledge of trigonometry (sufficient to know the meaning of sine, cosine and tangent). I have assumed that many readers will not be familiar with the calculus, and I have avoided using it—although it is possible that in so doing I may have made what is basically simple material appear more difficult than it is. The calculus is a branch of mathematics designed to solve problems involving quantities which vary continuously. For instance, the determination of the volume of a rectangular box is a matter for very elementary arithmetic, and so is the calculation of the volume of any solid which consists of a number of rectangular 'elements' of differing sizes. For a solid having a curved shape, however, there exists no simple arithmetical method of calculating the volume (although it would be a simple matter to measure it). Also, the speed of a car can be calculated by dividing the distance travelled by the time taken. But this will not give an accurate result if the speed is continually changing. In such a case, the calculus can be used to find the speed at any instant. In the examples where I have referred to the possible use of the calculus, it will be found that they concern situations where quantities (e.g. stress and radius) vary continuously.

Applied mechanics and mathematics contain a few examples of special words whose meaning might not be clear to a layman. Thus in Chapter Three the text refers to the 'clockwise' turning effect of a force. By this is meant that the force tends to turn an object in the same direction as the normal direction of motion of the hands of a clock. I have occasionally used the standard abbreviations for quantities (e.g. centimetres will sometimes be found as cm), but in general I have resisted the temptation to abbreviate and have written out quantities in full.

The sizes of physical quantities are often inconveniently large, or small, and for this reason are often multiplied or divided by multiples of 10. Most readers will be familiar with the mathematician's shorthand notation whereby 10 followed by a small number (called an index) represents the number of zeros after the 1. Thus 10^3 represents 1000, 10^6 is 1 million, and so on.

CHAPTER TWO

THE NATURE OF MOTION

Without concerning ourselves about the reasons *why* motion takes place (these we shall deal with in Chapter Four), let us consider a familiar case of a moving object: an aircraft crossing the Atlantic. The first consideration is where it is going to and where it has come from. In terms of applied mechanics, we call this the **displacement** of the aircraft. The second consideration (a very important one) is how long it takes to perform the journey. This is determined by the average rate of travel along the route chosen, and we call this the average **velocity**. Thirdly, and finally, we know that, although it is convenient for our schedules to speak of an average speed of, say 550 miles an hour, under practical conditions the actual speed is bound to vary from this. To take the two most obvious divergencies, the plane cannot *start* at 550 miles an hour, nor can it *finish* at this speed. The velocity, then, has to change from time to time, and the rate at which it does this is termed the **acceleration**. Let us look at each of these aspects of motion in turn.

(1) Displacement

Displacement seems to be a very simple concept. But if we are going to deal mathematically with it, which is exactly what the study of applied mechanics purports to do with physical situations, we have to be very careful to make sure that the rules of mathematics *work* when applied to each and every situation.

Let us try to apply a simple mathematical rule of addition to displacement. We know that, mathematically, 2 added to 3 makes 5. This is easy; but implicit in this simple statement are quite a few important conditions. First, the two things added must be of the same kind. Two years added to three months does not add up to five of anything. Secondly, applying the rule to displacement, it seems obvious at first that two feet added to three feet makes five feet; and indeed this is true if, for instance, we are measuring out material or merely measuring distance travelled. But displacement includes a little more than that.

Going back to our aircraft, we require a little more than that it shall travel 3200 miles from London. We want it definitely to finish up at New York. In other words, implicit in the displacement of a body is the **direction** it travels in. This being so, we now find that we cannot add displacements in the same way we can add sheep or molecules. Two miles due north added to three miles due west does not make five miles in any direction. If you want to know what it does add up to, you will have to draw a little sketch and either measure it or work it out. This has been done for you in Fig. 1.

Now there are three ways this little exercise could be performed in practice. First, you could walk two miles north, and then three miles west. Secondly, you could walk three miles west, and then two miles north. Thirdly, you could walk two miles north on *a body which was at the same time travelling three*

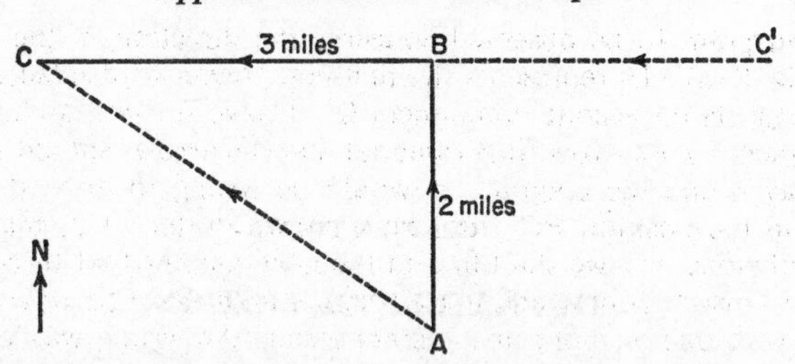

Fig. 1. Two miles due north added to three miles due west adds up to 3·6 miles in a direction 33·7° north of west. *AC* would be the actual path taken if the displacements occurred at the same time.

miles west. As an example of this last journey, you could sail two miles north in a sea which had a westerly current drift which took you three miles during your journey. You can see that, for the first two examples, the net displacement is the same (you can easily sketch the second journey for yourself). Whichever leg you walk along first, you will still finish at the same point, *C*. The same is true of the last journey, but in this case the line *AC* now represents the *actual path* taken by you. We shall have a good deal to do with these special additions, and we shall see that most of them deal with additions of concurrent quantities, like our two displacements in the third journey.

It is easy to extend this idea of adding displacements to more than two components. The simple rule for adding a number of concurrent or successive displacements is to draw a scale diagram of the various component displacements. The line joining the start of the diagram to the finish will then represent, to the same scale, the resultant displacement of all the components.

Fig. 2 shows two diagrams for determining the resultant of four displacements. There are two very important points to be noted in these diagrams. The first concerns the order in which the various component displacements are drawn, and it should be easy to see that the final resultant displacement is *not affected by this order*. At (*a*) and (*b*) I have shown two different orders of the components, the final resultant displacement *R* being the same. For this particular example, therefore, it would be possible to determine the resultant displacement by drawing any one of about 24 diagrams.

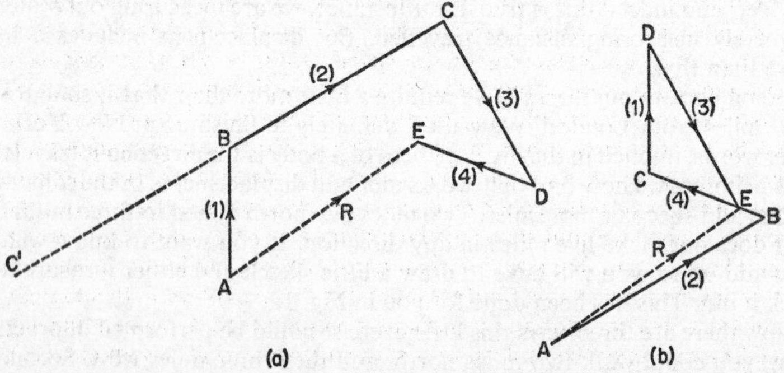

Fig. 2. The displacement vector diagram for four displacements. The resultant is the same no matter in what order the components are drawn.

The second point to be observed concerns the direction of the displacements. In Fig. 2 (*a*), *AB* represents the first component of displacement (1), and *BC* represents the second component (2). I have shown the direction of the displacement by an arrow. It is essential that the arrows should all follow the same way round the diagram. It would be wrong to draw the second component in the position BC^1, as I have shown dotted. This may appear simple and obvious, but we shall find in the next chapter that these rules of addition apply also to forces, and it is a very common error to draw diagrams of this type with one or more of the components the wrong way round. Of course, the arrow on the resultant *AE* will point from the starting point *A* to the finishing point *E*, and so will oppose the direction of the four components.

A simple example of four concurrent displacements might be the displacement of the tip of the rotor of a helicopter during a very short interval of time. The rotor spins relative to the hull, which gives one displacement. The hull moves relative to the ground, which gives the second component. The Earth spins about its axis, giving a third component. Finally, the motion of the Earth about the Sun gives a fourth. I should add hastily that this is an illustrative example only. No practising engineer is likely to spend his time in working out such an unlikely quantity as this. But it at least gives rise to the teasing question of what is the *true* displacement of the rotor, and the even more teasing answer that there is no such quantity. Displacements can only be stated relative to some point in space. For practical engineers, the point is usually a point on the Earth's surface, and for most cases this is assumed to be fixed. Another way of putting this is that, in calculations involving practically all motions of bodies on the Earth, the motion of the Earth itself can be ignored. It is, of course, a very different matter when a journey to the Moon is projected. The Sun then becomes the 'fixed' point.

Displacement is not the only quantity where we have to make use of special rules to perform addition. We shall find in Chapter Three that the same rules apply to force. Any quantity that is derived from displacement must also conform—as we are about to discover. Any quantity which requires these special rules of addition is called a **vector** quantity.

(2) Velocity and Speed

We have examined the nature of displacement in some detail. As a result, we can be reasonably brief in our examination of **velocity**, which is the rate of change of displacement. Miles per hour, kilometres per hour, feet per second, metres per second, are examples of the specification of velocity. Since the specification of displacement requires direction as well as magnitude, so must that of velocity. A velocity of 60 kilometres per hour is only half the story: the other half is the direction in which it takes place. The mere magnitude, irrespective of direction, is called the **speed**. The distinction is important because it means that, if the speed changes, the velocity must change, but the velocity may change without the speed changing. A car travelling east turns a corner and heads north, the speed remaining steady at 40 miles per hour, but its velocity has changed. As we shall see later, a force is involved in this change.

To find the effect of adding, or perhaps a better word is superimposing, velocities (for example, the motion of a ship subjected to an ocean current), we need merely consider displacement in a given interval of time. If we require the true 'earth' velocity of a ship whose ocean speed is 12 miles per hour due east in a current of speed 6 miles per hour due south, we simply draw a displacement diagram for a period of one hour, as in Fig. 3.

The answer is 13·4 miles per hour in a direction 26·5° south of east. But it must be realized that the diagram is essentially a velocity diagram and not a displacement diagram. It is merely convenient to take a period of an hour, but the diagram tells us the velocity *now*, at the instant, and gives no information about the state of motion one hour hence. One or other of the components may alter in the meantime. The velocity of a satellite relative to the solar axis may be found by drawing a velocity diagram compounding its velocity

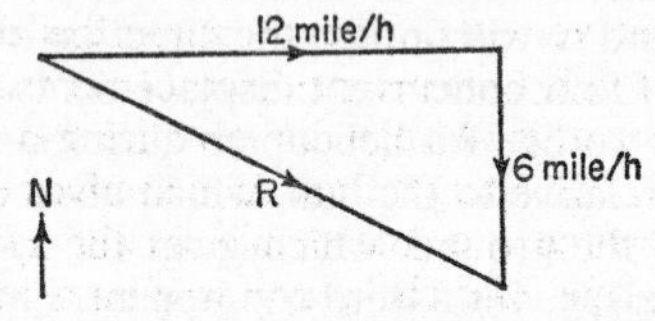

Fig. 3. The ship sails at 12 miles per hour due east in a southerly current of 6 miles per hour. The ship's true velocity relative to the land is *R* and is 13·4 miles per hour 26·5° south of east.

relative to Earth (say 18 000 miles per hour) and the velocity of the Earth relative to the Sun (say 70 000 miles per hour). The result will be the *instantaneous* velocity of the satellite, even though the diagram is drawn on the basis of displacement in one hour. But it can be appreciated that the two components, and hence the resultant, are changing every second.

(3) Acceleration

The rate of change of velocity is called acceleration. As such, it is a derivation of displacement, and thus the same rules of addition (that is, the rules of vector addition) apply when the resultant of a number of accelerations is required. The idea of acceleration as a change of speed is easy to comprehend. A motor-cyclist who starts from rest and achieves a speed of 70 miles per hour in 7 seconds may be said to have an acceleration of 70 miles per hour in 7 seconds, or, more simply, 10 miles per hour per second. But engineers usually frown on the practice of combining two different units of the same quantity, such as hours and seconds, and 10 miles per hour per second is more systematically expressed as 14·7 feet per second per second (usually abbreviated to 14·7 ft/s^2, or 14·7 ft s^{-2}). Nowadays, it is more likely to be expressed in metric units as 4·48 metres per second per second.

It is not quite so easy (although I hope not too difficult) to appreciate that acceleration occurs when the velocity changes in direction as well as magnitude. A car travelling at a constant speed around a bend is in a state of acceleration. The magnitude of this acceleration is determined by the speed of the car and the radius of the bend.

(4) Motion in a Straight Line

Although straight-line motion is the exception rather than the rule in applied mechanics, it forms a simple introduction to the determination of acceleration. This is because we can forget about the change of direction of velocity, and concentrate on the change of magnitude.

In Fig. 4 the symbols x, v, a and t represent displacement, velocity, acceleration and time, respectively. The three graphs in column (*a*) show how the first three of these quantities vary with time, for a condition of zero acceleration. It is seen that the velocity remains constant (as it must for this condition, by definition of the term acceleration) and that the displacement increases steadily, the displacement graph being a straight line. (In the terms of applied mechanics, the displacement is said to vary *linearly* with time.) An example of this type of motion would be that of a car travelling at steady speed along a straight level road.

In column (*b*) I have assumed a state of affairs in which the acceleration is constant. It is now the velocity which increases uniformly (and linearly) with time. If the acceleration is 4 metres per second per second, the velocity will be 16 metres per second after 4 seconds, will be 24 metres per second after 6 seconds, and so on. But since the speed is increasing all the time, it follows that the distance covered (i.e. the displacement x) will increase at a progressively-increasing rate, and the graph will be a curve of the form shown. This particular form of graph is called a parabola and will be familiar to many readers. It occurs here because it can be shown that the distance covered by the body is proportional, not directly to the time, as in column (*a*), but to the *square* of the time. Thus, if the moving object travels 4 metres

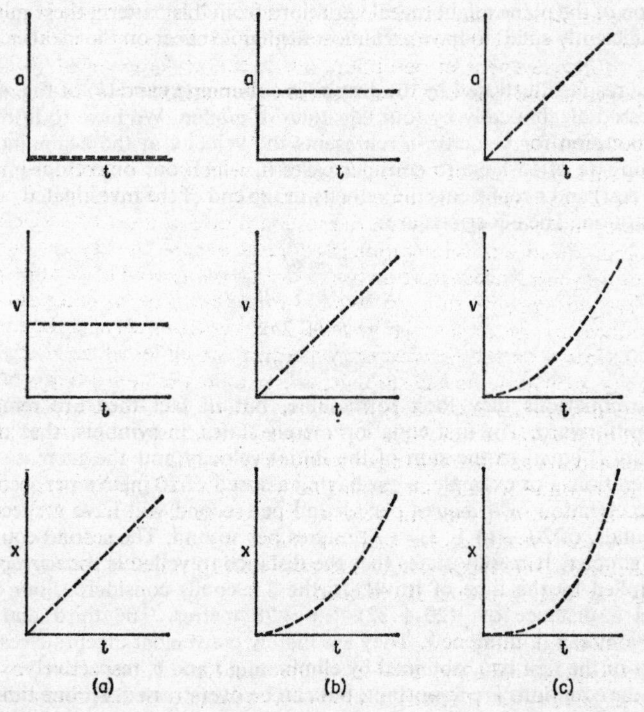

Fig. 4. (*a*) Motion with zero acceleration. (*b*) Motion with constant acceleration. (*c*) Motion with variable acceleration.

in the first second of motion, it will travel 16 metres (4×2^2) in the first two seconds, 36 metres (4×3^2) in the first three seconds, and so on. Examples of this type of motion are rare in practice; an approximation is found in the case of a freely falling body, if the resistance to motion of the air is neglected. This is why the trajectory of a cannon shell, or similar projectile fired upwards at an angle, has the approximate form of a parabola.

Column (*c*) of Fig. 4 shows the motion of a body under a steadily increasing acceleration. Now it is the turn of the velocity to increase parabolically, and the displacement increases at an even greater rate. There is no simple example of this special type of motion.

These three examples are merely chosen for the simplicity of the motion they illustrate. The actual motion of most real objects is usually extremely complex. But it is surprising how much can be achieved by considering even only the first two examples. It is possible, frequently, to make simple approximations, so that the motion may be *assumed* to be either constant acceleration or zero acceleration. Going back to our air liner *en route* between London and New York, the time of the journey could be determined with fair accuracy by assuming an initial stage of constant acceleration (while the craft got up to cruising speed), a second period of zero acceleration (at the cruising speed), and a final period of constant deceleration (negative acceleration) while the plane slowed down to a stop. Although the rigorous examination of the true motion of the plane might reveal variations from this pattern, these might well be sufficiently small to have an almost negligible effect on the ideal calculated time.

The results illustrated by the graphs in columns (*a*) and (*b*) of Fig. 4 can be expressed algebraically by four **equations of motion.** We have to introduce a new notation for velocity: u represents the velocity at the beginning of the motion (we often have to consider cases in which our object does not start from rest) and v represents the velocity at the end of the investigated period of the motion. The equations are:

$$v = u + at \tag{1}$$

$$x = \tfrac{1}{2}(u + v)t \tag{2}$$

$$v^2 = u^2 + 2ax \tag{3}$$

$$x = ut + \tfrac{1}{2}at^2 \tag{4}$$

These equations may look formidable, but in fact they are remarkably straightforward. The first equation merely states, in symbols, that the final velocity is equal to the sum of the initial velocity and the increase due to acceleration. For example, a car having a speed of 20 metres per second and an acceleration of 4 metres per second per second will have a speed, after 2 seconds, of $20 + (4 \times 3) = 32$ metres per second. The second equation is even simpler. It merely states that the distance travelled is the *average* speed multiplied by the time of travel. In the 3 seconds considered, our car will travel a distance of $\frac{1}{2}(20 + 32) \times 3 = 78$ metres. The third and fourth equations say nothing new. They are merely convenient algebraic rearrangements of the first two, obtained by eliminating t and v, respectively.

These equations are important, but can be overstressed. At one time it was the custom to drill students in the use of them *ad nauseam*, with the not-infrequent result that they could never forget them and would often use them in situations where they were not valid. Two examples of their use will suffice here. A car travelling along a straight level road passes a point A at 20 metres per second, accelerating uniformly, and passes point B 80 metres away in $2\frac{1}{2}$ seconds. What is its acceleration and final speed?

Here we are given the distance x, the time t, and the initial velocity u. To find the acceleration, we want the equation relating these terms and a. Equation (4) is the one to use. Substituting the given values:

$$80 = 20 \times 2\tfrac{1}{2} + \tfrac{1}{2}a(2\tfrac{1}{2})^2$$

Solution of the arithmetic gives $a = 9{\cdot}6$ metres per second per second. The final speed may now be obtained directly from equation (1). Alternatively it can be obtained from equation (2) without assuming the calculated value of a. Substituting in equation (1):

$$v = 20 + 9{\cdot}6 \times 2\tfrac{1}{2} = 44 \text{ metres per second}$$

For our second example, we need to know that the acceleration of a body falling freely to earth has a constant value (for most practical purposes) of 9·81 metres per second per second. A projectile is fired vertically upwards at an initial speed of 80 metres per second. What time will it take to reach a height of 200 metres? What will be its maximum height? And how long before it hits the ground?

Although this is straight-line motion, the object concerned reverses direction; we must therefore be careful to give all the quantities the correct signs. If we specify an upward direction as positive, then all downward quantities (displacement, velocity, and acceleration) must be specified as negative. With this in mind, the data given consist of initial velocity u, acceleration a, and for the first part of the question, distance x; the quantity required is time t. These four terms are included in equation (4). Substituting:

$$200 = 80 \times t + \tfrac{1}{2}(-9{\cdot}81)t^2$$

Readers who are familiar with algebra will recognize that this is a quadratic equation and can have two answers. Solving the equation, the two solutions are found to be $t = 3{\cdot}07$ seconds and $t = 13{\cdot}27$ seconds. The first answer is clearly the time taken to reach the stated height on the way up, and the second is the time at which the projectile passes the same point when on the way back down.

To determine the maximum height, we must consider that at this height the projectile is neither going up nor coming down: its velocity v must then be 0. Using equation (3):

$$0 = 80^2 + 2(-9{\cdot}81)x$$

giving a value of $x = 326$ metres.

For the last part of the question, we consider that when the projectile finally hits the ground it is back at its starting point: its displacement x is then zero. Using equation (4) again:

$$0 = 80t + \tfrac{1}{2}(-9{\cdot}81)t^2$$

This is another quadratic equation, and its two solutions are found to be $t = 0$ and $t = 16{\cdot}3$ seconds. The answer we require is clearly the latter, $t = 16{\cdot}3$ seconds.

(5) Angular Motion

Applied mechanics is concerned also with the motion of shafts, discs, electric motors, wheels and similar bodies which may spin but do not move bodily. The same terms of displacement, velocity and acceleration still apply. A shaft may turn through a given 'distance', it may have a certain speed, and its speed may increase or decrease. It merely becomes a problem of using a different concept of displacement.

Before we consider the choice of a unit for measuring angular displacement, let us look at the spinning disc illustrated in Fig. 5. Observe the motion of point P, which is situated at a radius r on the disc. There is obviously a con-

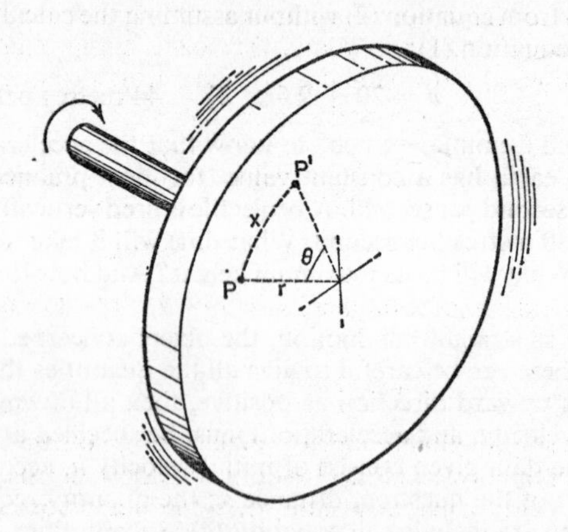

Fig. 5. While point P travels a distance x round the circle, the wheel turns through an angle θ.

nection between the angle through which the disc turns and the corresponding length of the path followed by P. For example, we know that if the disc makes one complete turn, P travels a distance equal to the circumference of the circle of radius r, so that the length of the path will be $2\pi r$. If the disc turns through only 1°, the corresponding distance moved by P will be $1/360 \times 2\pi r$. If it moves through 30°, the distance will be $30/360 \times 2\pi r$. So the distance x travelled by P for any angle $\theta°$ (theta degrees) turned by the disc is given by the formula:

$$x = \frac{\theta}{360} \times 2\pi r$$

For example, if a wheel of radius 0·8 metres turns through an angle of 20° a point on the circumference will move along a curved path of length

$$x = \frac{20}{360} \times 2\pi \times 0{\cdot}8 = 0{\cdot}279 \text{ metres}$$

If we now choose to express the angular movement in revolutions instead of degrees we obtain a different formula. For 1 revolution we calculated that the corresponding length of the curved path was $2\pi r$. By simple proportion, for N revolutions the distance x is given by

$$x = N \times 2\pi r$$

Using the same example as before, a rotation of 20° is a rotation of $\frac{1}{18}$ revolution. Thus

$$x = \frac{1}{18} \times 2\pi \times 0{\cdot}8 = 0{\cdot}279 \text{ metres}$$

Neither of these calculations is difficult. But they both contain the quantity 2π. We can avoid this if we adopt as our unit of angular displacement an angle of $1/2\pi$ times a full revolution. This angle is approximately 57·4° and we call it the **radian**. Now, suppose our disc turns through one radian, or $(360/2\pi)°$. The distance x covered by P will be:

$$x = \frac{\theta}{360} \times 2\pi r = \left(\frac{360}{2\pi}\right)\frac{1}{360} \times 2\pi r$$
$$= r \text{ metres}$$

If the disc turns through 2 radians, x will be $2r$, and so on. If the angle is θ radians:

$$x = r \times \theta$$

Thus, by choosing a suitable unit of angular displacement, we are able to obtain a very simple relationship between the linear distance travelled by a point moving along a circular path, and the corresponding angular displacement.

Now let us look at the velocity of our spinning body. To represent the angular velocity of spin we shall use the Greek letter ω (pronounced omega). (You must excuse the use of Greek letters, but they are so common in applied mechanics that to replace them would cause more confusion than it would remove.) The units in which ω is measured are radians per second. In Fig. 6 (*a*) we can observe the displacement of a point P in an interval of time t. If the linear velocity along the curved path is u, it will travel a distance along this path of ut; similarly, the corresponding angular displacement (in radians) will be ωt. But we know that $x = r \times \theta$. Therefore, $ut = r \times \omega t$. Hence:

$$u = r \times \omega$$

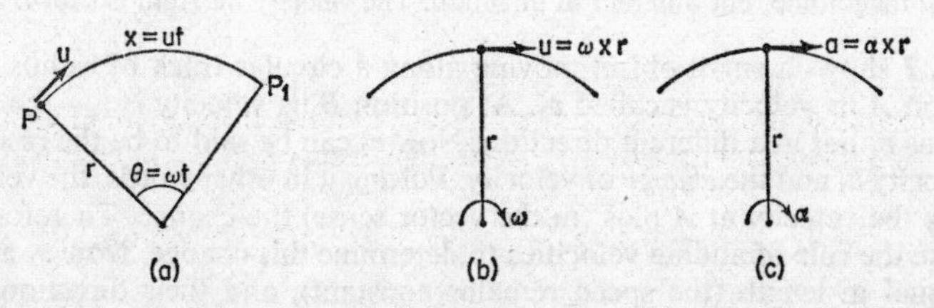

Fig. 6. The linear velocity is the product of angular velocity and radius. The linear acceleration is the product of angular acceleration and radius.

By a simple extension of the argument a corresponding relationship can be found between the linear acceleration a and the angular acceleration in radians per second per second, which is usually designated by α (alpha). The equation is:

$$a = r \times \alpha$$

We shall have occasion to return to these relationships when we examine the kinetics of angular motion in Chapter Four. Meanwhile, the motion itself can be analysed by a set of four equations which are the exact counterparts of those already stated for straight-line motion. Using the notation ω_1 and ω_2 for the initial and final angular velocities, the four equations are:

$$\omega_2 = \omega_1 + \alpha t \tag{1}$$

$$\theta = \tfrac{1}{2}(\omega_1 + \omega_2)t \tag{2}$$

$$\omega_2^2 = \omega_1^2 + 2\alpha\theta \tag{3}$$

$$\theta = \omega_1 t + \tfrac{1}{2}\alpha t^2 \tag{4}$$

Comparison of these equations with those on page 14 will reveal that they are in all respects exactly analogous.

(6) Motion in a Circular Path

Let us now examine the special case of the motion of a body which moves around a circular path at a constant speed which we will designate by v. We will call the path radius r. From what we have said concerning acceleration, you should now be able to appreciate that, although the speed of the object does not change, the direction of motion is constantly changing. In consequence, the body is in a state of acceleration. Let us see if we can find a method of determining this acceleration—probably one of the most important examples of acceleration in the study of dynamics. It is called **centripetal acceleration.**

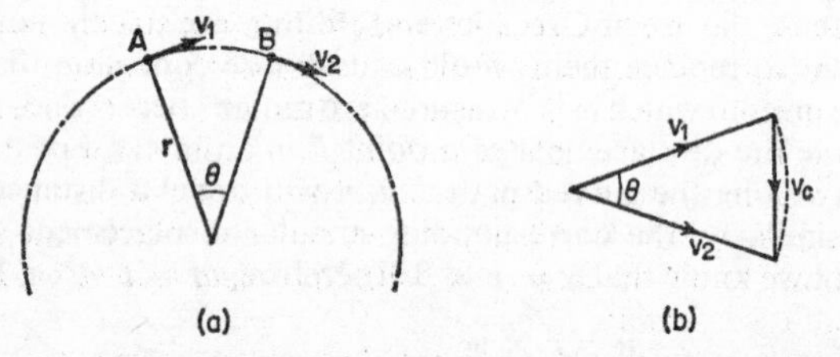

Fig. 7. In moving from A to B the velocity of the body changes from v_1 to v_2: the same in magnitude, but different in direction. The velocity diagram is shown at (*b*).

Fig. 7 shows a small object moving along a circular track of radius r. At position A its velocity is called v_1. At position B its velocity is v_2—the same value as v_1 but in a different direction. Now v_2 can be said to be the **resultant** of velocity v_1 and the *change* of velocity. Putting it in other words, the velocity at B is the velocity at A plus (in the vector sense) the change. Therefore we can use the rule of adding velocities to determine this change. Now v_1 and v_2 are equal in length (the speed remains constant), and their directions are tangents to the circle at points A and B, respectively. There is thus an angle θ between the two vectors equal to the angle between the radii at A and B. The change of velocity, which we can call v_c, must be the line completing the vector triangle shown in Fig. 7 (*b*).

If the angle θ is very small, we shall not introduce much error by saying:

$$v_c \simeq \theta \times v \quad \text{(where } \theta \text{ is in radians).}$$

The error is the difference of length between the base of the velocity triangle and the corresponding circular arc (shown by the dotted line).

Since θ is the angle between the radii at A and B, the distance x measured along the circular track is obtained from the formula

$$x = \theta \times r$$

(as we obtained in the previous section). Substituting in the first equation:

$$v_c \simeq v \times \frac{x}{r}$$

The acceleration is the rate of change of velocity, or v_c divided by the time taken to travel from A to B. Thus

$$a = \frac{v_c}{t} = v \times \frac{x}{rt} = \frac{v^2}{r}$$

because x/t is v, the speed along the path.

We have obtained this expression for the acceleration without satisfactorily dealing with the approximation we made. But you can see that we can make the velocity triangle with as small an angle as we like. In other words, we can make the calculation of the change of velocity for two points A and B *as close together as we choose to make them.* No matter how close we bring the points together, the reasoning still holds good and the error introduced by the approximation becomes less and less. Eventually, our two points become one, the error disappears, and we have the exact expression for the instantaneous acceleration of the body in its travel round the track. This is not the average acceleration over an interval of time, however small, but the acceleration at a certain instant, when the body is at a certain point on the track.

This may be clearer if we do a simple calculation. Take the case of a car travelling at a constant speed of 20 metres per second around a track of radius 100 metres, and let us determine the *average* acceleration as the car passes round angles of 30°, 10°, 5° and 1°, respectively. We can then compare the average acceleration in each case with the *instantaneous* acceleration determined from our formula.

Fig. 8 shows the diagrams of the motion, but with insufficient accuracy for

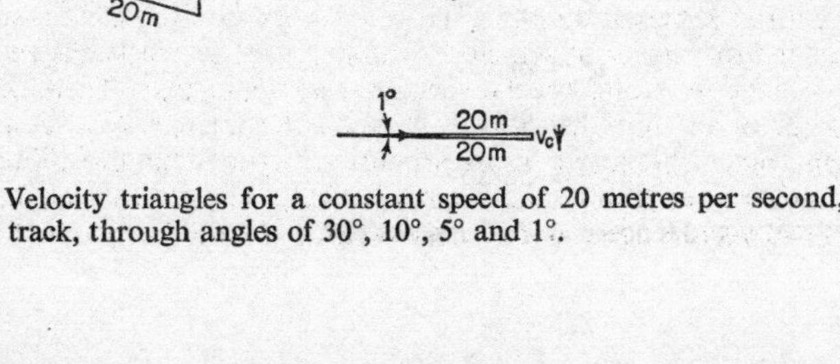

Fig. 8. Velocity triangles for a constant speed of 20 metres per second, along a circular track, through angles of 30°, 10°, 5° and 1°.

us to determine the change of velocity by measurement. Simple calculations using trigonometry give the respective values of v_c as shown in the table below. The actual distance round the track travelled by the car, and the corresponding actual time taken are also shown in the table. The last entry, the average acceleration, is obtained by dividing the velocity change v_c by this time.

Angle	30°	10°	5°	1°
Velocity change, v_c (metres/second)	10·352	3·488	1·744	0·349
Distance round track, x (metres)	52·36	17·45	8·73	1·746
Time, t (seconds)	2·618	0·873	0·437	0·0873
Average acceleration, v_c/t (m s^{-2})	3·955	3·990	3·995	3·999

The value given by our formula is

$$a = \frac{v^2}{r} = \frac{(20)^2}{100} = 4{\cdot}00 \text{ metres per second per second.}$$

The table shows that the value calculated from the triangle gets closer to this instantaneous value the smaller we take the interval of time, and even with an angle as large as 30°, the difference between the average and the instantaneous value is extremely small. It might be thought, since the formula always gives the same answer of 4·00 metres per second per second for the acceleration, that this is constant, and that the average acceleration over any fraction of the path should always be exactly equal to this value. But in fact the instantaneous acceleration is *not* constant. Although its magnitude remains unchanged, its direction is continually altering, as we are about to see.

Examination of the four velocity triangles of Fig. 8 shows that, as the angle gets smaller and smaller, the base angle of the isosceles triangle becomes closer and closer to 90°. In the instantaneous case, therefore, we can assume that the change of velocity (although we cannot actually draw it) becomes at right-angles to the velocity of the body. This is so. The acceleration of the body is at right-angles to its direction of motion at any instant, and this direction points always to the centre of the circular path. So, as the body swings round in its circular path, the acceleration vector swings round with it, and is *always directed towards the centre of the circle.*

By way of conclusion to this section, let us calculate the magnitude of this centripetal acceleration in a few simple cases. First, imagine yourself enclosed in a small cabin at the end of a swinging arm of length, say, 6 metres. At what speed would the arm be required to revolve in order to accelerate you at 9·81 metres per second per second (the constant acceleration of an object in free fall)? Substituting in the formula:

$$a = \frac{v^2}{r}$$

$$v^2 = ar$$

$$= 9{\cdot}81 \times 6$$

$$= 58{\cdot}86$$

$$\text{Hence } v = 7{\cdot}67 \text{ metres per second}$$

The rotational speed of the arm can be found by using the relation between linear and angular motion derived in Section 5:

$$v = \omega r$$

$$\omega = \frac{v}{r}$$

$$= \frac{7{\cdot}67}{6}$$

$$= 1{\cdot}28 \text{ radians per second}$$

which is easily converted to 0·204 revolutions a second, or 12·2 revolutions a minute—quite a moderate speed. The mechanical significance is that if you were to be swung at this speed in a vertical plane, the speed would be just sufficient to prevent you falling out at the top of the swing.

Let us now calculate the approximate centripetal acceleration of the Earth as it swings around the Sun. We will take the radius of the Earth's orbit as 93 million miles, giving a circumference of 585 million miles. As this takes approximately 365 days to complete, the mean speed (v) works out at 66 700 miles per hour, or 97 800 feet per second. (I am assuming that these astronomical details will be more familiar to most readers in the form given than in metric units.) The acceleration can now be calculated.

$$a = \frac{v^2}{r}$$

$$= \frac{(97\,800)^2}{93\,000\,000 \times 5280}$$

$$= 0{\cdot}0195 \text{ feet per second per second}$$

Finally, and as a contrast to this, consider the armature of an electric motor. Its diameter is 20 centimetres, say, and it runs at a speed of 3000 revolutions per minute—not an excessive speed for a motor. For this type of calculation it is more convenient to rearrange the formula by substituting for the velocity v straight away, thus:

$$a = \frac{v^2}{r}$$

$$\text{but } v = \omega r$$

$$\text{Hence } a = \frac{(\omega r)^2}{r} = \omega^2 r$$

In this example, ω is the rotational speed in radians per second

$$\omega = \frac{3000}{60} \times 2\pi = 314{\cdot}2 \text{ radians per second}$$

$$\text{Therefore } a = (314{\cdot}2)^2 \times 0{\cdot}1 = 9850 \text{ metres per second per second}$$

We shall find in Chapter Four that this sort of acceleration requires forces of a high order, which must be provided by the strength of the spinning armature itself. Any weakness in the armature may well result in its disintegration at high speed.

CHAPTER THREE

THE NATURE OF FORCE

There is probably no single engineering project in which the action of forces is not an important consideration. Consider for example a large structure such as a bridge. It is required to sustain the forces acting upon it due to the weights of objects passing over it, due to its own dead weight, and to the action of the wind on it. Furthermore, because of all these forces, the bridge itself transmits force to the earth upon which it rests, and the designer must ensure that the earth is either capable of withstanding these forces or can be suitably strengthened to do so. Piles are, in effect, vertical columns driven into the earth so that the weight of a large structure can be partly supported by forces of friction between the earth and the piles, and partly transmitted to harder earth or rock lower down. The designer of an engine must ensure that the forces generated by combustion of gas can be safely contained within a cylinder, and that the forces which are used to create high-speed rotary or reciprocating motion are not sufficient to cause the engine to fly to pieces.

One of the first things we have to do is to arrive at some sort of definition of force. We must state with accuracy what it is and what it does, and we must agree on a method of evaluating it: that is to say, a system of units. Now a rigorous definition of force is the province of **dynamics**, which we shall not encounter until the next chapter. Our definition must accordingly be provisional, and to that extent unsatisfactory, but it will suffice for us to find out quite a lot of useful information.

It helps that most of us really know when force is encountered, even when we cannot define it accurately. We know that a cart is *pulled* by a horse, that a train is *pulled* by the engine, that the Earth is *pulled* by the Sun, that a piston is *pushed* by gas or steam pressure, that your toe is *pushed* by the weight of someone's foot in a crowded vehicle. We can accept then, as a provisional definition, that force is a *push* or *pull* exerted by one object on another.

Most of us know also that force is measured by a scale of weight. To say that an object weighs 2 tons is another way of saying that the Earth is pulling it downwards with a certain force, the magnitude of which we can judge by experience. When we are told the magnitude of forces in various circumstances, such as the thrust on a rocket, the force on a car-engine piston, the weight of a molecule, or the breaking strength of nylon rope, we usually try, often with indifferent success, to relate the information to forces within our own experience, such as the weight of our own bodies, or the weight of a car.

I have already indicated in Chapter One that the unit of force has been standardized as the newton. Leaving aside the definition of this unit, it is sufficient for our present purposes to accept that the newton is 0·225 of the weight of one pound, or approximately one-tenth of the weight of one kilogramme. For forces of much greater magnitude, we use the familiar prefixes kilo- and mega- to denote one thousand and one million, respectively.

Another convenient point of reference is that ten kilonewtons is approximately one ton.

Force usually manifests itself in one of two very distinct and different situations, which must be readily distinguishable to the student. First, a number of forces may be present and may act on an object in such a way that the combined effect upon the *motion* of the object is zero, although it is vital to appreciate that the object itself is affected by the forces. A typical example is the bridge mentioned in the opening paragraph of this chapter. As a result of all the forces acting, the bridge does not (we hope) move bodily and continuously, although the material of the bridge is deformed and stressed due to the forces. In such a case, all the forces are said to be in equilibrium, and the problem is one of **statics.**

The second situation we encounter is that in which the combined effect of forces upon the motion of the body is not zero, and a change of the pattern of motion results. A simple example is the propulsion of a rocket. Although many forces may act on the object, they do *not* all cancel out, as they do on the bridge, and the stationary rocket moves, or the moving rocket changes its direction. This type of situation is a problem of **dynamics.**

As I have already indicated, we shall study dynamics in Chapter Four, and the remainder of this chapter will be devoted to the laws of statics. But you must not suppose that engineering problems always fall conveniently into such neat classifications. We shall find that a great deal of what we study in this chapter will require application in the next. In fact, statics is better thought of as the study of forces themselves, without regard to their effect, than as the study of stationary objects.

(1) Identification of Forces

Since force is so important, our next step is to make sure we can identify force when it is encountered: this is not so easy as might be thought. If the effect of forces is to be determined, it is absolutely necessary that *all* the forces concerned are taken into account. It would be a disaster for a rocket designer to calculate for a journey to the Moon, and forget the *weight* of the rocket; or for the designer of a supersonic air liner to forget the resistance of the air. Let us start by classifying our forces into three categories: direct forces, remote forces and induced forces.

Direct forces are the most obvious, and thus the easiest to identify. They exist by virtue of the direct contact between two bodies: the pressure of a *real* gas on a piston, the friction of a *real* road surface on a wheel, the thrust of a *real* jet effluent on an aircraft.

Remote forces are not so obvious, but are so limited in type that they are usually easy to spot. These are the forces that act at a distance, and the most familiar example is that of gravitation. You should clearly distinguish between this type of force and a direct force. It is not necessary for the Earth to *touch* a body in order to pull it. The only other forces in this category are magnetic and electrostatic forces.

Induced forces are those that arise *because of the action of direct or remote forces.* A lift cage is pulled downwards by the remote force of gravity, and it is pulled *upwards* by the tension in the supporting cable. If the force of gravity were removed, the tension would disappear. A train remains at rest on rails because the downward weight is exactly balanced by an upward reaction—the thrust upwards of the rails on the wheels. Similarly, a ship floats on the sea because of the upward thrust of the water on the hull of the ship, exactly balancing the downwards weight. Remove the one, and the other will disappear. Make the ship lighter, and it will float higher, resulting in less upward thrust. The tension of the supporting cable, the upward reaction of rail on wheels, and the upward thrust of the water on the ship are all examples of induced forces.

It may help towards the understanding of these induced forces if we consider the question often asked by elementary students in this field. A crate is lowered on to a floor, and immediately there is an upward reaction by the floor on the body, which disappears as soon as the crate is lifted again. The rather naïve, but perfectly admissible, question is how does the floor know when or when not to push? If our floor is replaced by a platform supported on springs, perhaps the question answers itself. We can see that the lowering of a load on to the platform causes a *compression of the springs*, and this compression will cease as soon as the load set up in the springs exactly equals the load being lowered. This feature is true (with modifications) for any induced load. The induced load is created by deformation of the supporting body, whether this is the ground, a rope, the sea, or a stone wall. We shall find in Chapter Eight that all materials deform to some extent under even the smallest load.

(2) The Free-body Diagram

Most problems of applied mechanics resolve into the study of forces on one object, or body, as it is usually called, and a strict discipline is needed to ensure that all forces are taken into account. The body is taken (in imagination) and hung freely in space; then *all* the forces acting on it are considered, one by one, and marked on a diagram of the body, which is called a free-body diagram. First, all remote forces can be shown. If the problem is earth-bound, there is usually only the weight to consider, and this always acts downwards. The establishment of direct and induced forces is helped by the knowledge that these can only exist by virtue of the body's touching something outside itself. In technical language, these forces arise by virtue of the body's **constraints.** Ordinary common sense helps a great deal here. If the body rests upon a surface, it may require some effort to imagine the surface pushing upwards (which it does), but imagination boggles at the idea of the surface pulling the body downwards (which it cannot possibly do). Similarly, a rope supporting a load can conceivably pull upwards, but under no circumstances can a flexible rope push. If frictional forces are known to exist, they will always act in the direction opposite to the motion, or to the direction of incipient motion if no real motion exists.

Fig. 9 shows three examples of bodies subjected to forces, and the associated free-body diagrams. First we have the case of an aircraft in flight. The weight W acts vertically downwards. Also, the aircraft is pushed forwards by the propeller thrust, which is shown as T. The two remaining forces are due to the contact of the aircraft with the air. The forward motion of the aircraft causes the air to exert an upwards force (called lift) on the wings and tail-plane; this is indicated by L. In addition, the friction of the air causes a retarding backward force called drag; this is shown as D. At a constant cruising speed, the engine thrust must exactly balance the drag, and the lift must exactly balance the weight. The shape of the cross-section of the aerodynamic surfaces of an aeroplane (i.e. the wings and tail-plane) is called an aerofoil, and this section is specifically designed to give as great a lift as possible, for as small a drag as possible, under flight conditions.

Fig. 9 (*b*) shows a diagrammatic version of a simple lift or hoist, consisting of the lift cage, a counterbalance weight, a connecting rope and a pulley. Both cage and counterweight are each subjected to weight (downwards) and rope tension (upwards) only, if the frictional resistance of the air is ignored. The air resistance, which was one of the major forces in the first example, is here so small as not to seriously affect the problem. The two rope tensions have been

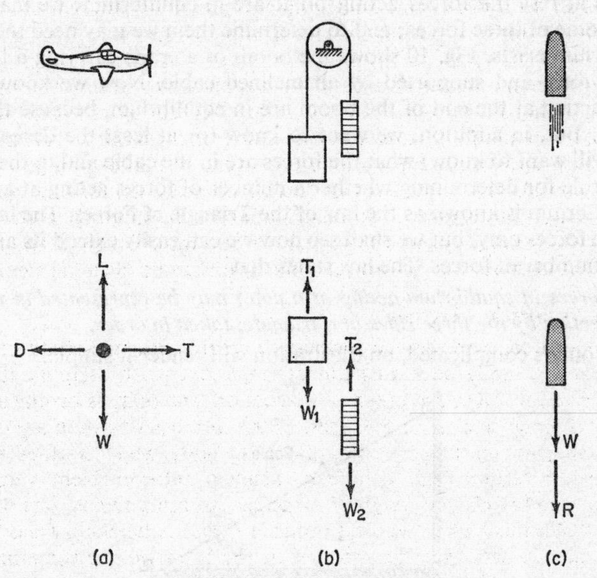

Fig. 9. Three examples of bodies subjected to forces, and the free-body diagrams: (*a*) an aircraft in flight; (*b*) a lift cage with counterweight; (*c*) a bullet in vertical flight.

labelled T_1 and T_2, respectively. It is tempting to ascribe a value to these tensions, equal to the weights they oppose, but to do so would be to assume more than one knows at this stage. We shall find in the next chapter that many cases arise when the tension in a supporting rope is not equal to the weight hanging on it.

Fig. 9 (*c*) illustrates a bullet fired vertically upwards. A brief consideration discloses only two possible forces: the weight W, and the resistance of the air R, both acting downwards. This example is included to illustrate a very common failing found among students, which is to deduce the existence of a force *because the bullet is in high-speed flight*. I cannot stress too strongly that there is no justification for such a deduction, which cuts right across the principles formulated in the preceding paragraphs. The real explanation is that the bullet was set into high-speed motion by forces within the barrel of the gun, and these ceased to operate the instant the bullet left the barrel. At the identical instant, the *bullet began to slow down* owing to the action of the two real forces we have shown—forces which will eventually bring the bullet to rest, and one of which (the weight) will cause it to reverse its direction and fall back to earth.

(3) The Triangle of Forces

Earlier in this chapter I have referred to the state of equilibrium. It is now our business to arrive at some rules for determining whether forces are, or are not, in equilibrium—rules which must be independent of the body upon which the forces may be acting. It is not sufficient to observe that because a body is at rest the forces acting on it are in equilibrium: we may need to know some of these forces; and to determine them we may need the fact that equilibrium exists. Fig. 10 shows the boom of a crane carrying a load of 10 kilonewtons, and supported by an inclined cable. Now we know that the forces acting at the end of the boom are in equilibrium, because there is no motion. But, in addition, we want to know (or at least the designer of the crane will want to know) what the forces are in the cable and in the boom.

The rule for determining whether a number of forces acting at a point are in equilibrium is known as the law of the **Triangle of Forces**. The law applies to three forces only, but we shall see how we can easily extend its application to any number of forces. The law states that

Three forces in equilibrium acting at a point may be represented in magnitude and direction by the three sides of a triangle, taken in order.

If this sounds complicated, an illustration will render it simple.

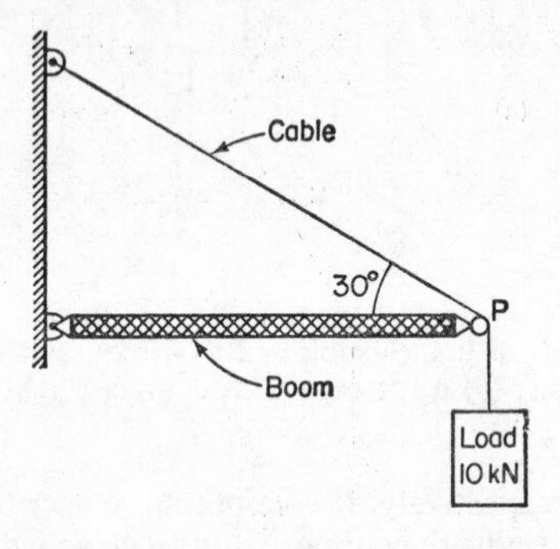

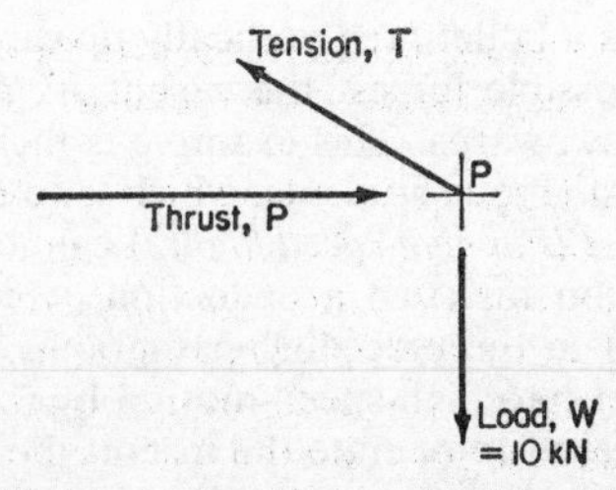

Fig. 10. Point P is in equilibrium under the action of the three forces shown.

Suppose we have a system of three weights and cords, and two smooth pulleys, as shown in Fig. 11. Let us examine the junction of the three strings, *P*. Three forces act at this point, namely the three tensions in the cords. Although I have warned about equating a string tension to the weight it supports, you may safely do this in this case. So the three tensions may be assumed to be equal to the weights of 2, 3·5 and 4 kilogrammes, respectively.

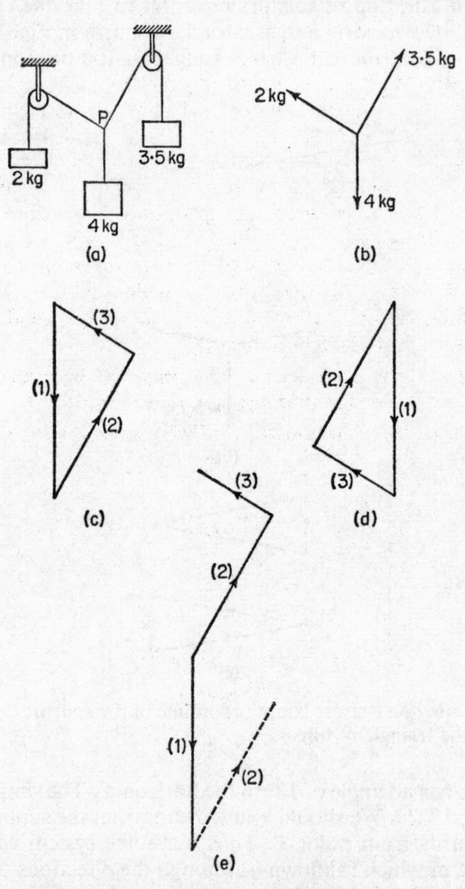

Fig. 11. The point *P* is subjected to three forces equal to the respective string tensions. Vectors representing these forces form a closed triangle.

The simple force system is shown in Fig. 11 (*b*). Now the triangle of forces law states that, if we draw lines parallel to each force in turn, of length proportional to the magnitude of the force, the three lines so drawn will form the closed sides of a triangle. This has been done in Fig. 11 (*c*) and also in Fig. 11 (*d*), the difference being the order in which the forces were drawn. If you have attentively read Chapter Two you will by now have recognized our old friend the vector, which we used for displacements and velocities. The warning I issued concerning the direction of the arrows around the figure applies here also. In Fig. 11 (*e*) I have shown the result of failing to conform to this rule: the figure does not form a closed triangle because vector (2) was drawn with the wrong 'sense', i.e. with the arrow opposing the direction of the arrow on vector (1). Vector (2) should have been drawn in the dotted position.

As a simple illustration of this principle, let us take the case of a street-lamp suspended by two wires across a road, as shown in Fig. 12.

Let us suppose that the lamp has a weight of 100 newtons, and that the

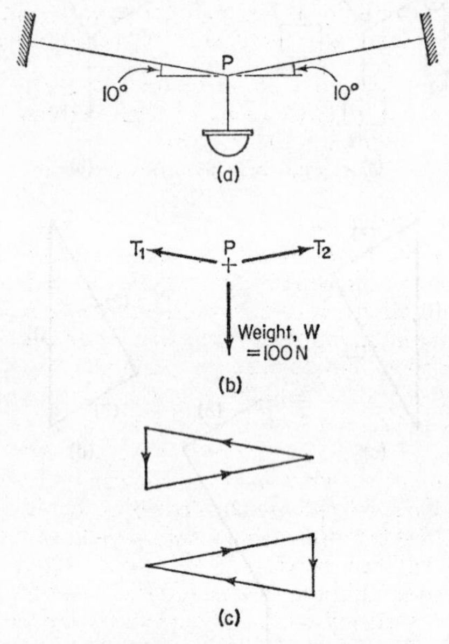

Fig. 12. Forces acting on a street-lamp: (*a*) outline of the system; (*b*) the three forces acting at *P*; (*c*) the triangle of forces.

supporting wire has an angle of 10° to the horizontal. The simple force system is shown in Fig. 12 (*b*). We should know by now that the supporting rope can only pull outwards from point *P*. Thus our force system consists of three forces, only one of which is known—although the directions of the other two are known. We start our triangle by drawing our known downward force of 100 newtons. This we may do by drawing a vertical line, say 10 millimetres long, with the arrow pointing downwards. Now we may either draw a line from the bottom of this, parallel to T_1, and from the top parallel to T_2; or we can draw T_2 from the bottom, and T_1 from the top. In either case, the *directions* of the two forces must be observed, and the line drawn in the corresponding direction. Both alternatives have been drawn in Fig. 12 (*c*). Notice that we do not need to know the length of the two lines if we know their directions; in fact, we measure these lengths after the triangle is completed, and thus determine the values of the rope tensions. In this case, both lines are found to be approximately 29 mm in length, which represents a force of 290 newtons to the scale we have chosen.

Before leaving this little example, it is interesting to speculate on the effect of reducing the angle of depression of the supporting wire. Without introducing further calculation, it can be seen that the effect will be to elongate the triangle, so that the oblique sides become flatter, and proportionally longer. If, for instance, we make the wire absolutely horizontal, we finish up with a triangle whose long sides never meet, indicating that the resulting tension would be infinite. Such a situation would be impossible to achieve in practice, for no matter how tight the wire, the central load must cause it to sag a small amount. Even if there is no central load, the dead weight of the rope or wire itself must cause a small amount of sag. This is illustrated by the jingle:

> 'No force on earth, however great,
> Can stretch a string, however fine,
> Into a horizontal line
> That shall be absolutely straight.'

(4) Resultant and Equilibrant

I hope that by now you can see that the triangle of forces enables us not only to evaluate forces in equilibrium, but also to examine systems of forces which are not in equilibrium. If only two forces act at a point, they cannot be in equilibrium (except in the special case where they are exactly equal and act in opposite directions). We may use the triangle of forces to find the **resultant** of the two forces, in just the same way that we found the resultant of a number of displacements in Chapter Two. If the two forces are represented by the two sides of a triangle, the third side represents the resultant. However, the direction of the resultant is determined by drawing the arrow in the *opposite* direction (around the triangle) to the arrows on the two components. If the arrow on the third side is drawn in the *same* direction as the two components, this third vector will then represent a force which, if added to the two existing ones, will produce equilibrium. Such a force is called the **equilibrant**, and it is easily seen that it is exactly equal and opposite to the resultant.

I have illustrated this in Fig. 13 (*a*), where F_1 and F_2 are the two forces. In Fig. 13 (*b*) these are represented by the vectors (1) and (2). The closing side of the triangle is E, representing force F_E which is the equilibrant. In Fig. 13 (*c*) the triangle is shown with the closing side having the arrow pointing in the opposite direction, representing force F_R which is the resultant of F_1 and F_2.

The determination of the resultant of a number of forces is probably one of the most important aspects of statics. The resultant may be defined as that

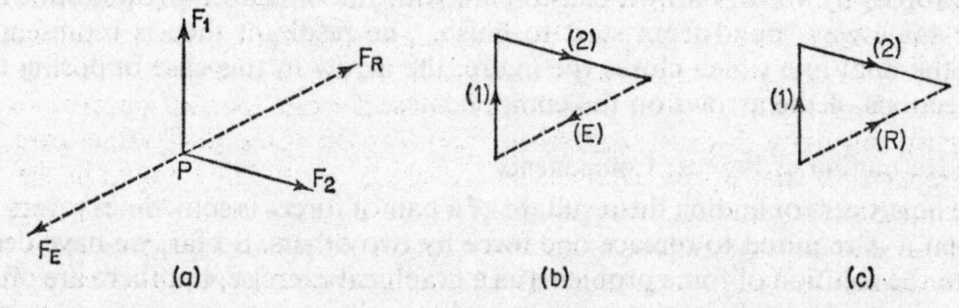

Fig. 13. Only two forces act at P. F_E is the equilibrant; F_R is the resultant.

force which, acting singly, produces exactly the same effect as the component forces. We must now develop this technique to find the resultant of more than two forces.

(5) The Polygon of Forces

One fears proportionate complication when we increase the number of forces; fortunately the added complication is one of number, not of principle. Look at the five forces shown in Fig. 14. Let us assume that they are in equilibrium. Now we know that if we have only three forces, we can prove equilibrium by drawing the triangle of forces. But we now have a method of reducing two forces to one, by obtaining the resultant. Let us do this with

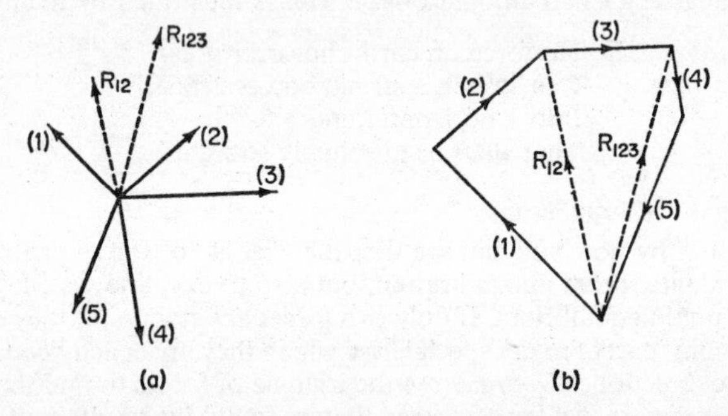

Fig. 14. The polygon of forces. Forces (1) to (5) may be represented by the sides of a closed polygon.

forces (1) and (2). In Fig. 14 (*b*) the resultant of these forces, R_{12} is shown as the closing line of the triangle. Note that the arrow opposes the direction of the two component forces. If we include R_{12} in the force system, as shown dotted in Fig. 14 (*a*), we can mentally delete (1) and (2) because the resultant replaces them.

So we are down to only four forces. But we can repeat the process, this time with the two forces R_{12} and (3), giving us a second resultant, R_{123}. This can be coupled with (4) and so on. What emerges is, of course, a polygon comprising vectors representing the five component forces. If these forces are in equilibrium, the polygon must close. What is more, there is no need to draw the various resultants at all. Thus, for any number of forces in equilibrium, acting at a point, a closed polygon of forces may be drawn. For a number of forces not in equilibrium, the resultant may be found by representing the forces by vectors drawn end-to-end, with the direction arrows following the same way round from start to finish. The resultant then is represented by the final line which closes the figure, the arrow in this case opposing the directions of the arrows on the components.

(6) Resolution of Forces: Components

The process of finding the resultant of a pair of forces is sometimes reversed, when it is required to replace one force by two others. So far, we have dealt with the solution of force problems as a graphical exercise, but there are often occasions when this is inconvenient and laborious.

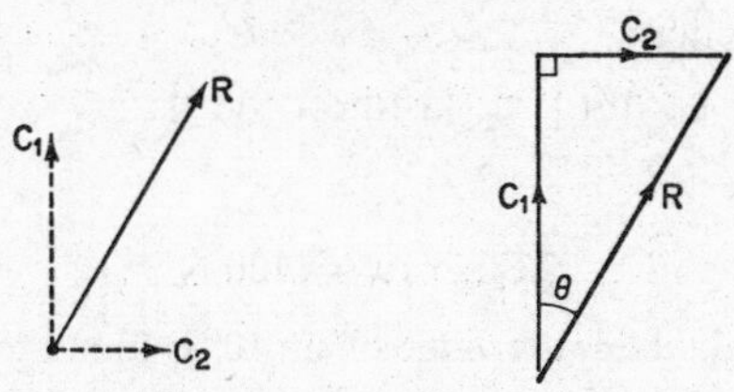

Fig. 15. Force R can be replaced by its components C_1 and C_2.

Examine the situation shown in Fig. 15. We have a single force R acting at a point. I have drawn a force triangle alongside, in which R is the resultant of a pair of forces C_1 and C_2, and I have assumed these to be respectively vertical and horizontal. So the single force R can be replaced by its two **components** C_1 and C_2. The advantage is that, if we assume them at right-angles, as I have done, we can see from the simple trigonometry of the right-angled triangle that C_1 is $R \cos \theta$ and C_2 is $R \sin \theta$. It is therefore no longer necessary to draw the triangle at all. We can use the street-lamp example from Section 3 to illustrate this method of solution.

Fig. 16 (*a*) shows the original force system, consisting of the weight, 100 newtons, and the two tensions, T_1 and T_2. The tensions have been replaced in Fig. 16 (*b*) by their respective components, $T_1 \sin \theta$, $T_1 \cos \theta$, $T_2 \sin \theta$ and $T_2 \cos \theta$. This process is called **resolving** the force. So we have in Fig. 16 (*b*)

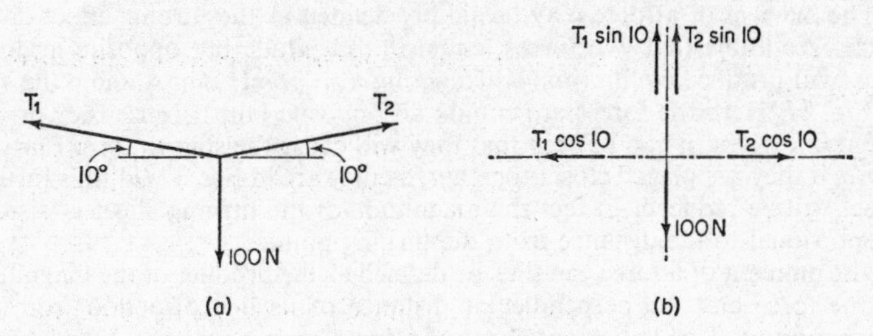

Fig. 16. Forces T_1 and T_2 are resolved into components.

a neat system of forces acting only along two lines at right-angles. The advantage is that we can write equilibrium conditions along each direction independently of the other. Equilibrium along the horizontal is unaffected by forces along the vertical. In other words, *a force can have no component along a direction at right-angles to its line of action.* If you tow a truck along a rail by walking alongside, pulling a rope attached to the truck, you can only cause it to move if the rope makes an angle of less than 90° with the rail, so that the tension of the rope has a forward component. If the rope is exactly perpendicular to the rail, no amount of pulling will move the truck along the rail; it will merely pull it over.

In the present example, we can write an equation of equilibrium along the horizontal direction:

$$T_1 \cos 10^\circ = T_2 \cos 10^\circ$$

from which T_1 clearly equals T_2.

In the vertical direction:

$$T_1 \sin 10° + T_2 \sin 10° = 100 \text{ N}$$

But since $T_1 = T_2$:

$$2T_1 \sin 10° = 100 \text{ N}$$

From mathematical tables, the value of sin 10° is found to be 0·1736. Hence

$$T_1 = \frac{100}{2 \times 0{\cdot}1736} = 288 \text{ newtons}$$

The solution is easier, and the answer more accurate, than our previous solution using the triangle of forces.

Structural engineers make use of this method to determine the forces in the members of frame structures such as simple bridge girders. The method is particularly suitable because the directions of the members of such a framework are usually horizontal or vertical or inclined at simple angles such as 30° and 45°. We shall examine this branch of statics in some detail in Chapter Seven. You can probably use this method to prove for yourself that the cable tension T of the crane boom of Fig. 10 is 20 kilonewtons and that the thrust P in the boom is 17·32 kilonewtons.

(7) Moment of a Force

The **moment** of a force may be simply defined as the turning effect of the force. We know that two forces, equal in magnitude but opposite in direction, will produce equilibrium when acting at a point; but examine the ship in Fig. 17. Here the forces are equal and opposite but, because they do not act at one point, it can be seen that they will cause the ship to turn. Furthermore, if they are placed closer together, as they are in Fig. 17 (*b*), this turning effect will be reduced. In fact the magnitude of the turning effect is directly proportional to the distance from the turning point.

The moment of a force can thus be defined as the product of the magnitude of the force and the perpendicular distance of its line of action from the turning point. This important effect of a force must now be included in our conditions of equilibrium. For equilibrium to be assured, not only must a

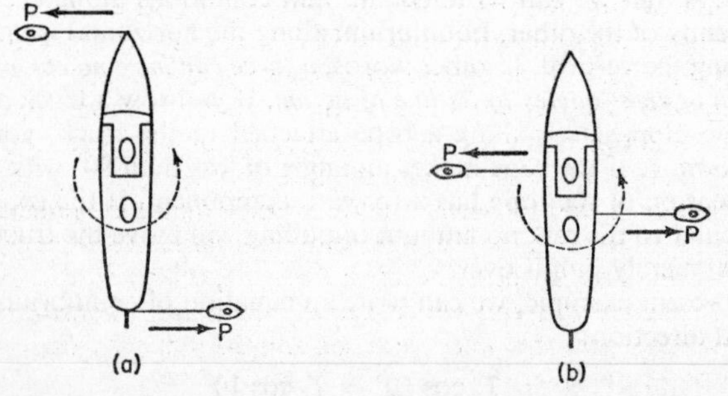

Fig. 17. The forces are equal and opposite, but do not act at one point. The result is a turning effect on the ship.

system of forces have no resultant, but also the moments of all the forces must total to zero. Furthermore, this must be true for *any* turning point.

We shall first examine one or two simple cases of moments in which the 'turning point' I have referred to is obvious: cases in which the system actually does turn about a specific point. Fig. 18 shows a simple lever arrangement for lifting a weight. The lever is hinged to a fixed support at its left-hand end, and force is applied at some other point along the lever. In such a case, our obvious turning point is the hinge. In order to lift the weight, it is necessary that the moment of the force we apply about the hinge must over-balance the moment

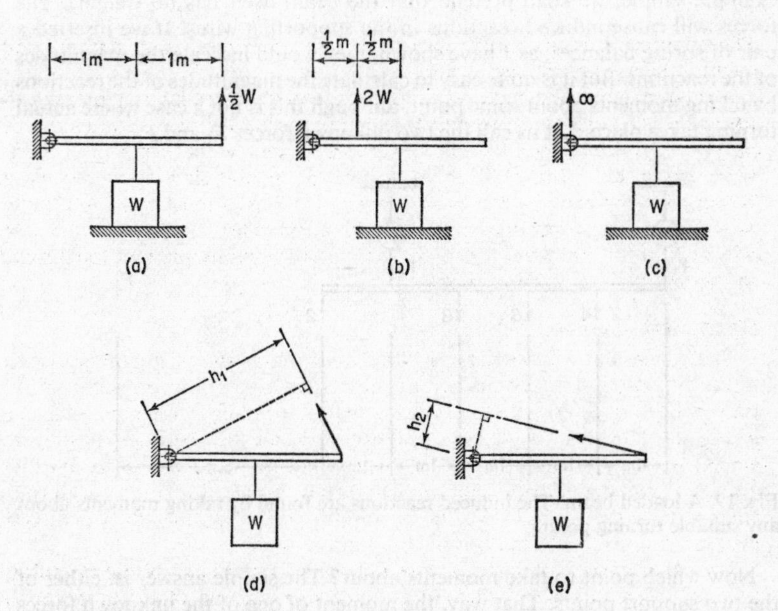

Fig. 18. The moment of a force about a hinge is the product of force and perpendicular distance from line of action to the hinge.

of the weight about the same point. By applying the force a distance from the hinge twice as far as the weight, as in Fig. 18 (*a*), we find that the necessary force is $\frac{1}{2}W$, where W is the weight. But with the arrangement shown in Fig. 18 (*b*), the force required is $2W$. In Fig. 18 (*c*) I have shown the force applied at the hinge: it has a moment of zero *about* the hinge, and no amount of force will cause the weight to be raised.

In passing, note that the weight has a moment about the hinge at all times. Why, then, does it not turn? Because, until it is raised from the table, there is an induced reaction of the table on the weight, which is exactly equal and opposite to the weight, and which has an equal and opposite moment. As soon as our force acts, the weight is raised and the induced reaction is replaced by the force we apply.

In Figs. 18 (*d*) and (*e*) I have shown the force in the same position as in Fig. 18 (*a*), but this time it is not perpendicular to the beam. In calculating the necessary force we have to take moments about the hinge; if you recall the definition of 'moment', you will realize that we have to take the product of force and *perpendicular* distance from hinge to line of action. These perpendicular distances are h_1 and h_2. As the distance gets smaller, the force must be greater to exert the same moment.

Now let us look at a case where there is no obvious turning point. The horizontal beam in Fig. 19 supports forces of 4, 6, 8 and 2 units (say kilonewtons) in the positions shown. It is hung from two points. The forces can be assumed to be the weights of masses attached to the beam (to keep the example simple, we shall pretend that the beam itself has no weight). The forces will cause induced reactions in the supporting wires. If we inserted a pair of spring balances, as I have shown, they would indicate the magnitudes of the reactions. But it is quite easy to calculate the magnitudes of the reactions by taking moments about some point, although this is not a case where actual turning takes place. Let us call the two unknown forces F_1 and F_2.

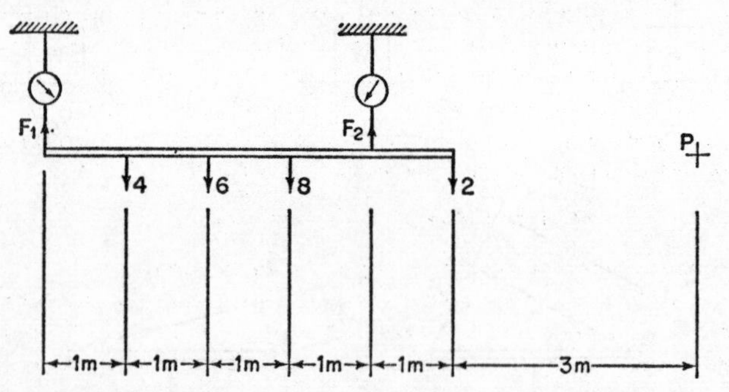

Fig. 19. A loaded beam. The induced reactions are found by taking moments about any suitable turning point.

Now which point to take moments about? The simple answer is, either of the two support points. That way, the moment of one of the unknown forces becomes zero; it is thus conveniently set aside, leaving us to calculate the other. But it must be stressed that this answer is one of expediency. In theory, we could take moments of all forces about *any* point. If we did, we should have an equation which would contain *both* of our unknown forces F_1 and F_2, and anyone familiar with elementary algebra knows that you cannot determine two unknown quantities with only one equation. So we would have to take moments about *any other* point, and obtain a second equation. Our pair of simultaneous equations could then be solved for F_1 and F_2 by standard methods. I am not going to do this, because the work is made so much simpler by a careful choice of the point about which you decide to take your moments. This choice is one of the things that students of applied mechanics have to learn, and practitioners have to know. In examples like this one the choice is easy, but sometimes a wrong choice can lead to a great deal of unnecessary calculation. For our present example, let us choose the left-hand support as our 'turning point'.

You can see that, if the left-hand support is regarded as a possible hinge, the four downward-acting forces will all tend to turn the beam the same way (clockwise) but the upward pull of the reaction F_2 will have an opposite turning effect. This must be taken into account in our equation. We shall call the clockwise moments positive, and the anti-clockwise negative. Since the beam is in equilibrium, we know that the *total* moment must be zero. The equation is thus:

$$(4 \times 1) + (6 \times 2) + (8 \times 3) + (2 \times 5) - (F_2 \times 4) = 0$$

from which simple calculation gives $F_2 = 12\frac{1}{2}$ kilonewtons.

For the sake of the exercise, let us take our turning point as the right-hand support. Now, the upward-acting force F_1 and the force of 2 units both exert a clockwise moment. All remaining moments are anti-clockwise. This gives our second equation thus:

$$(F_1 \times 4) + (2 \times 1) - (8 \times 1) - (6 \times 2) - (4 \times 3) = 0$$

showing F_1 to be $7\frac{1}{2}$ kilonewtons.

But we have done almost twice as much work as necessary. Since the whole beam is in equilibrium, we know that there can be no resultant *force* upwards or downwards. This by itself is not sufficient information to find F_1 and F_2, but having found one of these, we can use this fact to find the other. Thus, having found F_2, we write a simple equation of force in the vertical direction, calling downward-acting forces positive and upward negative.

$$4 + 6 + 8 + 2 - F_1 - F_2 = 0$$

and since F_2 is known to be $12\frac{1}{2}$, F_1 is seen to be $7\frac{1}{2}$ as before.

To show that if equilibrium exists, moments about *any* point total to zero, let us take a point P which is 3 metres to the right of the end of the beam. Taking due account of the directions of the moments, our equation is:

$$\begin{aligned}\text{Moment about } P &= (F_1 \times 8) + (F_2 \times 4) - (4 \times 7) - (6 \times 6) - \\ &\qquad (8 \times 5) - (2 \times 3) \\ &= (7\tfrac{1}{2} \times 8) + (12\tfrac{1}{2} \times 4) - (4 \times 7) - (6 \times 6) - \\ &\qquad (8 \times 5) - (2 \times 3) \\ &= 60 + 50 - 28 - 36 - 40 - 6 \\ &= 110 - 110 \\ &= 0\end{aligned}$$

proving our earlier assertion.

(8) Centre of Gravity

This is a rather elaborate name for the point of balance. Real objects occupy a definite amount of space. They are not concentrated at one point. But it is possible to pick them up at a certain point so that there is no tipping as a result. The beam we examined in Fig. 19 was supported at two points. Let us now imagine it to be supported at one point only. It is obvious that if we merely remove the support at F_2, the beam will tip round until weights and beam all hang vertically below the other support. A similar result would attend the removal of the left-hand support. But you can see that it should be possible to support the beam at a single point such that, when picked up at this point, the beam would stay balanced horizontally. This support point is called the **centre of gravity**. Its position can be found experimentally by trial and error, but it can also be calculated by writing an equation of moment equilibrium. Let us draw the beam again, this time showing only one support, and as we do not know where it is, we shall let it be a distance x from the left-hand end.

For this sort of problem, deciding which point to adopt as our turning point is easy: any point will do. Let us choose the point P that we used for our earlier example. Now the upward supporting force, which we will call F,

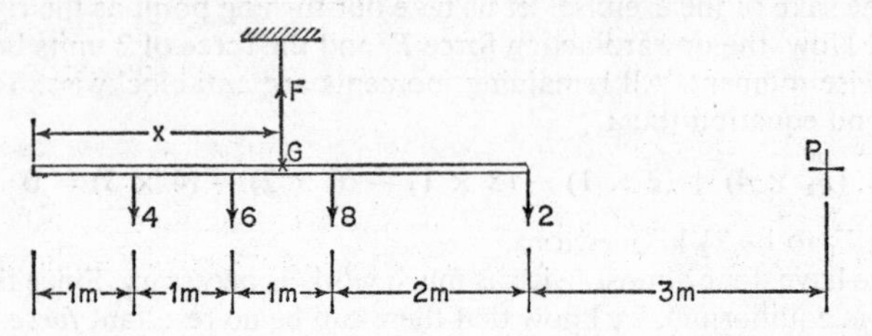

Fig. 20. G is the centre of gravity. Its position may be determined by taking moments of forces about some point.

must balance the total downward force. We do not need an equation this time to tell us that this total force is 20 kilonewtons. So we can now write our equation of moments about point P, taking due account of directions as before. This time, all the forces are known, but the distance x is not known. The equation is:

$$F(8 - x) - (4 \times 7) - (6 \times 6) - (8 \times 5) - (2 \times 3) = 0$$

$$20(8 - x) = 28 + 36 + 40 + 6$$

$$= 110$$

$$8 - x = 5\tfrac{1}{2}$$

$$x = 2\tfrac{1}{2} \text{ metres}$$

If you wish, you can obtain the same result by writing an equation choosing any other point as turning point. For instance, if you take the left-hand end of the beam as turning point, you will have a simpler equation, but the same result.

The centre of gravity is of great importance, both practically and theoretically. In engineering workshops and factories, when objects frequently have to be lifted by cranes, the problem of finding the point of balance may often be dodged to some extent by picking up at two points, using wire slings. But sometimes this method may not be practicable. For instance, large electric motors or generators might be badly damaged by slinging them by the two shaft-ends. In such cases, manufacturers often fit a lifting ring screwed into the framework of the machine exactly over the centre of gravity.

The calculation of the position of the centre of gravity for objects in which the weight is distributed continuously (as distinct from being distributed in chunks as we used for our example) may be very simple, or it may be extremely complex. As an example of a simple case, take that of a uniform beam or shaft. The point of balance is clearly half way along. Similarly, for flat uniform sheets of metal, of simple regular shape such as rectangles or circles, the centre of gravity is along the axis of symmetry. The centre of gravity of quite irregular shapes can often be found approximately by dividing them up into a series of simple regular rectangles and circles. The method of the calculus can be used to locate the centre of gravity of shapes which vary continuously. An example would be a tapering shaft.

The distribution of the load in a freight-carrying aircraft is a practical exercise in centre of gravity. Distributing the load too much to the front or rear might seriously affect the trim of the plane in flight. The considerable weight of fuel in the tanks is also an important factor, and the trim of the plane may need adjustment over long flights to allow for the shift of centre of gravity as the fuel is used up.

Engineering handbooks often contain exhaustive tables of standard shapes and sections, with the position of the centre of gravity indicated, to save busy engineers the task of calculating it. But most busy engineers usually find that the particular case they want happens to be one not included in the table.

(9) Torque

A particular case of moment of a force or forces arises where the result of such a moment is a continuous turning or a continuous twist in one direction. This occurs when shafts are used to transmit force or power. A shaft is a very useful and convenient method of transmitting the action of a force over long distances. In such cases, the word 'moment' is seldom used. The product of force and distance from shaft centre is called the **torque**, and it is measured in newton metres (N m). This is really a matter of convenience. The shaft of an electric motor is caused to turn by a magnetic force applied to the armature of the motor. The product of the magnetic force and the radius of the armature will constitute the torque of the machine. For an engineer who is going to use the motor, neither the value of the magnetic force nor the size of the armature is of any importance: all he is concerned with is the magnitude of the twist that the shaft will provide him with, and the speed with which it will run. If he is going to use the motor to drive, say a hoist, part of his job will be to calculate what torque is required to operate the hoist. Let us consider a very elementary, almost crude, example.

Fig. 21 shows a very simple type of hoist. The weight is carried by a rope which passes over a pulley, and this in turn is driven by a motor connected to its axis. Suppose the weight of 10 kilonewtons is to be raised a height of 10 metres in 20 seconds, and say that the pulley has a diameter of 0·4 metres. What torque and speed will be required of the driving motor?

The torque is the product of force and radius; in this case, $10\,000 \times 0{\cdot}2 = 2000$ newton metres. To find the motor speed, we must recall some of the work of Chapter Two. The speed v of the weight is related to the speed ω of the motor by the formula $v = \omega r$. In this case, $v = 10/20 = 0{\cdot}5$ metres per second. Hence:

$$0{\cdot}5 = \omega \times 0{\cdot}2$$

giving a value of $2\frac{1}{2}$ radians per second for the angular velocity, ω. This can be easily changed to revolutions per minute, giving a value of 24 revs per minute.

I intimated that this was a crude example. No electrical engineer is likely to supply a motor with this torque and speed. The torque is far too high and the speed is far too low. In real hoists, there would almost invariably be a gear-box connected between the motor and the pulley. This would have the dual effect of permitting the motor speed to be higher, and of allowing the motor torque to be lower. Machines such as electric motors, petrol engines, diesel engines, and steam turbines, work most efficiently and economically when turning at relatively high speeds. The purpose of the gear-box is to reduce this speed to a convenient one for the job in hand, and as I have indicated, and as I shall show in more detail in Chapter Five when I come to talk about power, the 'side effect' is that the torque is increased in the same ratio that the speed is reduced.

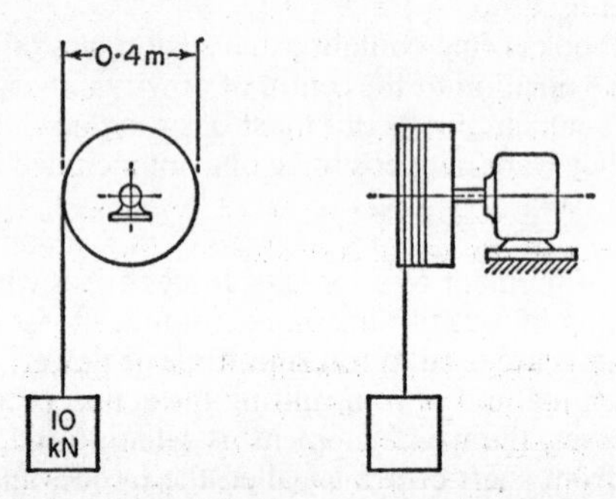

Fig. 21. A simple hoist: the weight is supported by the rope passing round the pulley; the pulley is turned by the motor.

This explains why a car engine can be stalled easily by trying to start off in a high gear. The load of the car is transmitted to the engine as a torque. In top gear, the speed ratio of engine to wheels is the lowest ratio—that is to say, the car has the highest possible speed for a given engine speed, but is driven by the lowest torque. The highest torque is transmitted in bottom gear, at the lowest speed. Now it always requires more torque to start a car than to keep it running; if top gear is engaged, the required starting torque is higher than the engine can provide. Consequently, the engine stalls. If, on the other hand, you use the car to start the engine (as many of us have to do, alas), you may find the reverse situation. Connecting in bottom gear requires such a high torque at the wheels that the weight of the car on quite a steep hill may be insufficient to turn the engine. The simple remedy is to connect in top gear, or maybe third. Bear this in mind next time your friend asks you to give him a push.

CHAPTER FOUR

THE LINK BETWEEN FORCE AND MOTION

We have examined motion in Chapter Two without inquiring into the reasons why motion changes. In Chapter Three we have briefly studied force in its most general aspect, but have not considered the effects of forces. A considerable part of the remainder of this book will deal with the effects of forces in various situations. Chapters Eight to Ten will deal with the behaviour of some engineering materials when subjected to forces. In Chapters Eleven and Twelve we shall examine the forces encountered in static and moving fluids. But this chapter is concerned with one of the most fundamental aspects of dynamics—the effect of forces on movable bodies.

(1) Newton's Laws

Unlike statics, the findings of this branch of mechanics are comparatively modern, dating only from the 17th Century and the time of Galileo and Isaac Newton. This is not the place to write the history of this work: we are merely concerned with the results, which here can be expressed in Newton's famous Laws of Motion. Since Newton wrote his work in Latin, we are permitted a degree of flexibility in rewriting these laws (this is why you may often see them expressed in different forms). Let us first state the laws and then examine them singly.

1. Every body continues in a state of rest, or of uniform motion in a straight line, unless or until acted upon by a force.
2. The acceleration of a body is proportional to the force acting upon it, and takes place along the line of action of the said force.
3. Every force induces an equal and opposite reaction.

The first law gives us a definition of force: force is something which, by itself, produces an acceleration. This is *the* definition of force, and is clearly more satisfactory than the 'push or pull' we have had to accept so far. It only gives us a qualitative definition: it does not tell us how to *measure* force, but we shall find the answer to this problem in the second law.

Let us now see how general the first law is in its application, and how it must have appeared to cut right across contemporary beliefs. It is natural to assume that a body which is not acted upon by a force would be in a state of rest: it is not so obvious to assume a possible state of motion in a straight line. All terrestrial experience of Newton's day must have pointed to other conclusions. All bodies on the Earth, if set moving, came 'naturally' to rest. On the other hand, bodies apparently free from earthly interference (planets, for instance) were known to move in approximately circular paths. From the time of the Greeks, one school of thought accepted the 'natural' motion of bodies as circular. We now know that earth-bound bodies are subjected to frictional force, and that planets are subjected to gravitational force: both of these forces cause departure from straight-line uniform motion. In fact, perhaps the most exasperating aspect of Newton's theories is that no body ever observed has a 'natural' motion unaffected by force. There is an intriguing and amusing dialogue on this topic between Newton and the artist, Kneller in G. B. Shaw's play, 'In Good King Charles's Golden Days'. It is perhaps typical of Newton's genius that he was able to perceive, without the benefit of direct observation, what the unforced motion of a body would be. Present-day students have less difficulty with this idea than those of an earlier age. Most schoolboys now appreciate that astronauts, once on course for the Moon, can just sit back (perhaps 'sit' is the wrong word here) and let the straight-line motion carry them at thousands of miles an hour to their destination without the expenditure of any fuel. Of course this is an approximation—the capsule is subjected to gravity pulls from Earth, Moon and Sun—but the basic idea that, once set on its journey, a craft needs no further force to propel it, is now fairly widely appreciated.

Henceforward, then, we have to think of force as that which produces an acceleration—or that which would do so, if acting alone. The qualification is important. There are many instances of forces which do not produce acceleration, but every one of these forms part of a system where forces are in equilibrium and cancel each other out. Acceleration is produced only by a single force acting alone, or by the **resultant** of a number of forces. Thus, a car is pushed forwards by the tractive force applied at the wheels. It will accelerate only if this tractive force exceeds the resisting force due to road friction and air resistance. When the two forces are equal, the car reaches a steady speed. The increase of the resisting force of the air with increase of speed is the factor which sets a practical limit to the speeds of planes and cars. (Extraterrestrial vehicles, having no air to contend with, are not subject to this limitation.)

When a body is known to be subjected to a constant, sustained single force, we can expect a constant and sustained acceleration. The classic instance of this is planetary motion. The approximate circular motion of a planet around the Sun is *produced* by the gravitational pull of the Sun. If the latter were suddenly to be removed, the planet would fly outwards along a tangential path. Observe the difference between uniform motion in a circular path, and uniform motion in a straight line. We found in Chapter Two that the former constituted an acceleration, which is directed towards the centre of the circular path.

Newton's Second Law re-states the first, but adds the vital word 'proportional'. This law, above all others, is based on experiment, mostly undertaken by Galileo. If the first law helps us to *understand* dynamics, it is the second law that helps us to *use* it.

The third law is perhaps the most subtle and the most difficult to understand. It states that whenever a force exists, there must be an equal and opposite force in the same system. This is *not* the same as stating that equilibrium must exist, as I hope to show. Equilibrium exists when the resultant of forces acting *on a single body* is zero. If you hang from a rope, the downward force of gravity is balanced by an upward force due to the tension in the rope. This tension is exactly equal and opposite to the weight. But this is *not* the equal and opposite reaction referred to in the third law. It cannot be, because merely by cutting the rope we establish a situation whereby the downward weight continues to act, but the upward tension no longer does so. Where, now, is the 'equal and opposite reaction'?

In order to understand this fully, you must appreciate that force *always acts between two bodies*, and a true understanding of the nature of force is only possible when you examine it in the context of *both* the bodies concerned. I realize that in Chapter Three I have stressed the importance of the free-body diagram when considering forces in a system, but this is only a device, useful for simplifying the determination of the resultant force, and does not really contribute towards the understanding of force itself. At this stage, I shall take the liberty of restating Newton's Third Law in more explicit terms:

> 'When a force exists between two separate bodies *A* and *B* in a system, the force exerted upon *B* by *A* is exactly equal to the force exerted upon *A* by *B* but opposite in direction.'

Now let us go back to you, still hanging from your rope. The first thing to look for is body *B* (assuming that you are content to be body *A*). Since the force concerned is that of gravity, body *B* must be the Earth. Are we to assume, then, that Newton's Third Law makes the absurd assertion that the Earth is being pulled upwards by your body? It is indeed, and by the same force as your weight. Unfortunately, the Earth is so big that we tend to lose sight of it as an independent body capable of being acted upon by, and responding to, forces. The situation becomes clearer if we make a simple model of a gravitational system between two bodies. Imagine two rowing boats floating freely on a lake. We can represent gravity by tying a rope to boat *A*, and pulling on it from boat *B*. Granted that the pull will not vary with the distance, as the pull of gravity does, but that does not matter for our purpose. Now, the question to ask is, 'Which boat moves towards the other?'

For those readers who unhesitatingly reply that boat *A* moves towards the puller in boat *B*, I recommend the method invariably adopted through the ages by true scientists—go and try it! You will find that *both boats move towards each other*, and will eventually meet at some point between them. If you now transfer yourself to boat *A* and pull boat *B*, the difference will be very slight. In a few words: as you pull boat *B*, boat *B* pulls you with an equal and opposite force. Of course, you really know this without thinking about it. Suppose your rope is tied to the jetty instead of to the other boat, which way do you pull if you want to move towards the jetty? Away from the jetty, of course, knowing that the jetty, via the rope, will pull you in the opposite direction to your pull.

As to the question of how far each body moves, the answer is that they move through distances inversely related to their weights. If the two boats are of exactly equal weights, they will meet exactly at the middle of the line joining them. If one boat is replaced by a Thames barge, the barge will move perhaps one inch for each five yards moved by the small boat. If the barge is replaced by an ocean liner, it becomes clear that any movement there might be is for all practical purposes unmeasurable. When the ocean liner becomes the Earth, we even forget that it is a body at all.

I could quote examples by the score, but you should think them out for yourself. I have quoted gravitation, but any force between two bodies has the same property. A locomotive pushes a truck along a straight level track. The third law states that the truck pushes backwards on the locomotive. If you are so misguided as to insert your hand between the buffer plates while this is taking place, you can argue the philosophical point with your doctor afterwards: was the force on the back of your hand, or the front? Obviously, if you had been blindfolded, you would have had no means of telling which was the pushing locomotive and which was the pushed truck. All you can assert is that there was a force *between* the two. So far as the truck alone is concerned, there is a forward push, and the locomotive by itself suffers a backward push.

The next objection is that if the truck pushes the locomotive backwards, why does the latter move forwards? Well, of course, if this were the only force on the locomotive, it would actually move backwards. But the locomotive also pushes itself backwards against the rails, by a force transmitted to the wheels from the pistons, which is at least as great as the push forwards on the truck. The locomotive moves forwards due to the forward push of the rails on the locomotive, which is the equal and opposite reaction of the push of the locomotive on the rails. Obviously, the truck cannot be pushed with a force greater than the tractive thrust of the locomotive on the rails, and if this force is reduced (say by water, or ice) the locomotive cannot push the truck, or even itself, forwards.

This duality of a force is the essence of dynamics. The locomotive pushes the truck forwards; the truck pushes the locomotive backwards. The cannon forces the shell forwards; the shell pushes the cannon backwards. The jet engine thrusts the effluent backwards; the effluent pushes the engine forwards. The Earth pulls the Moon; the Moon pulls the Earth. But when we come to examine the forces acting *on a single body*, we have to cut across this duality. Because the free-body diagram is essentially a map of forces on a single body, it can never show the equal and opposite reaction which must be acting on the *other* body of the system. If, for example, we analyse the motion of the railway truck, the relevant force is the single forward push by the locomotive. We must not include the force exerted *by* the truck *on* the locomotive. We take this latter force into consideration when we come to examine the motion of the locomotive.

(2) The Second Law and Gravitational Acceleration

Although we have spent some time in examining the third law, you will realize that it contains nothing which helps us to make a quantitative use of the relationship between force and acceleration. This is the province of the second law, which is based fundamentally on experimental work carried out by Galileo. The idea of **proportionality** is simple, and is no doubt understood by most readers, but it is so important that I must make it absolutely clear.

Let us think of a few examples of quantities which are mutually proportional. If we travel in a car at a constant speed of 10 metres a second, we can state with confidence that the distance we travel is proportional to the time taken: thus in twice the time we shall travel twice the distance, and so on. If we hang weights on a spring balance, the movement of the pointer is proportional to the weight; the divisions of weight on the balance scale are equally spaced. Contrast this scale with the divisions on, say, a pendulum-type kitchen scale, which will be found to get closer and closer together towards the heavy end of the scale. The movement of mercury in a thermometer is proportional to the increase of temperature. Examples of quantities which are related but *not* proportional include: the volume of a cube and the length of its side, the distance covered by a freely falling body and the time taken, the volume of a quantity of gas in a cylinder and its pressure, the weight and the pointer movement on the pendulum-type scale mentioned above. In Fig. 22 I have shown what the graphs of all these related quantities would look like.

The graphical evidence of proportionality is a straight line; indeed, the relationship between the quantities is said to be a **linear** relationship. When

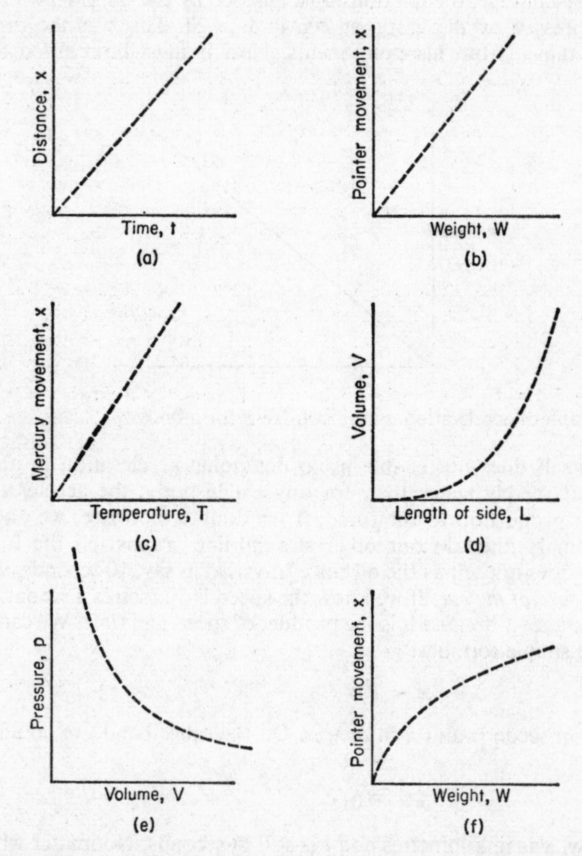

Fig. 22.
Graphs showing related quantities: (*a*)–(*c*) are proportional, (*d*)–(*f*) are not.

(*a*) Distance against time for a car travelling at constant speed.
(*b*) Pointer movement of a spring balance against weight.
(*c*) Mercury movement of a thermometer against temperature.
(*d*) Volume of a cube against length of side.
(*e*) Pressure of a quantity of gas against volume, temperature remaining constant.
(*f*) Pointer movement against weight for a pendulum-type scale.

the relationship between the quantities is not linear, it is often possible to find a relationship which is. For example, instead of Fig. 22 (*d*) we could draw a graph of volume V against (length of side, $x)^3$; we should then obtain a straight line, as you can easily check for yourself. In Fig. 22 (*f*), we could obtain a straight-line graph by plotting the pressure against the *reciprocal* of the volume, $1/V$, illustrating the famous law of Robert Boyle for gases at constant temperatures. This device of finding linear relationships between quantities is very useful in applied mechanics, particularly in experimental work.

In the present case, the relationship expressed by the second law of motion can be expressed by the graph shown in Fig. 23. This was the result that Galileo obtained from his experiments. Now it must be realized that this

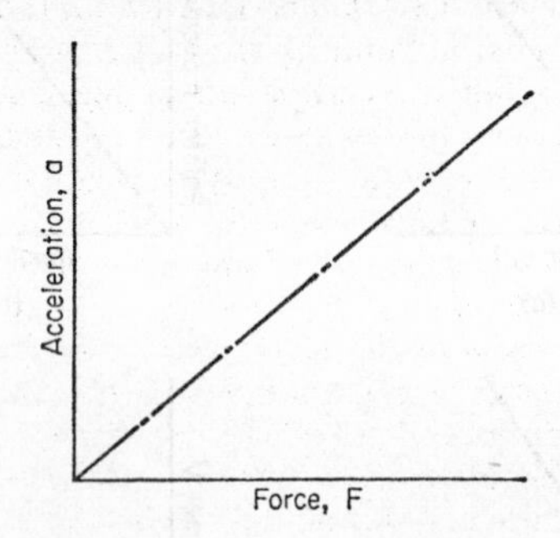

Fig. 23. Graph of acceleration a against force F for a body.

graph in itself does not enable us to determine acceleration in any given situation. It merely states that, for any single body, the acceleration will increase in proportion to the force; if we double the force, we double the acceleration. Neither do our other straight-line graphs tell the full story. Fig. 22 (*a*) does not tell us the distance travelled in say, 10 seconds, *unless we know the speed of the car*. If we know the speed is 10 metres a second, we can find the distance x because it is the product of speed and time. We can express this by the simple formula:

$$x = 10t$$

where t is in seconds and x in metres. On the other hand, we could use the formula:

$$x = 0{\cdot}01t$$

where, now, x is in kilometres and t is still in seconds. No matter what units we choose, the pattern of the law is always the same:

$$x = Ct$$

where C is some number whose value is determined by the conditions; in this example, the speed of the car, and the units we choose.

Now the second law of motion may be expressed by a similar formula:

$$F = Ca \qquad (1)$$

which means that the relation between force and acceleration for any single body can be obtained by multiplying a by a number whose value is determined from the conditions of the experiment and the units chosen.

(3) Mass and Weight

How do we set about finding the value of this number C? Anyone who has done simple graphical work will know that you can draw a straight-line graph of the type shown if you know just one point on the graph or one experimental result. Going back to Fig. 22 (*a*), our distance/time graph for the car, if we know that the distance x is 45 metres when the time t is 3 seconds, we can provide the graph with numbered scales, *and* we can find the speed of the car.

In the matter of force and acceleration, we have an advantage. It is a known fact that all bodies subjected only to the pull of gravity at any given place on the Earth's surface fall at the same rate of acceleration. This was proved experimentally by Galileo (although the story of his dropping heavy bodies from the Leaning Tower of Pisa is almost certainly not founded on fact). If bodies do not fall with the same acceleration, the reason is that they are subjected to forces other than the pull of gravity—usually the resistance of the air. The resulting acceleration due to gravity has been determined with high accuracy in various parts of the world and at various altitudes. Although slight variations do occur, owing to irregularities in the spherical shape of the Earth and to other factors, it is sufficiently accurate for most purposes to assume that the acceleration due to gravity is constant. The value usually assumed is 9·81 metres per second per second (32·2 feet per second per second), and this quantity is denoted algebraically by the letter g.

So, for any single given body falling freely to earth, we may be assumed to know the force acting upon it (its weight) and also its resultant acceleration (g). Denoting the weight by W, we may now substitute W for F, and g for a in equation (1) thus:

$$F = Ca$$

Hence $$W = Cg$$

and $$C = \frac{W}{g}$$

We now have a value for the constant number C for a body, providing we know its weight. Substituting this value back into our equation (1), we get the Equation of Motion for a body:

$$F = \frac{W}{g} a \qquad (2)$$

It is of course a known fact that W, the weight of the body, varies from point to point in space, and (theoretically at least) can be zero. How, then, can the quantity W/g be a constant of proportionality for a body? The answer is that as W varies from place to place, g varies in exactly the same proportion. If our body is transported to the Moon, its weight will then be approximately one-fifth its weight on Earth. But the value of g_M, the Moon's gravitational acceleration, will also be approximately one-fifth the value of g on Earth. In other words, bodies will fall to the Moon's surface at a rate of acceleration approximately one-fifth of the rate they fall on Earth, or approximately at 2 metres per second per second.

This number which we have called C, and which we can now evaluate for any particular body (assuming we can weigh it), represents a measure of the resistance of the body to a force. The bigger the C, the smaller the acceleration for a given force. It is a number expressing quantitatively the **inertia** of the body. Unlike the weight of the body, which varies from place to place, inertia is constant. At least it is constant for a given system of measurement. If we choose to change our measurement of acceleration from metres per second per second to feet per second per second, the number C will of course be different. Apart from this, the value of the number can only be altered by alteration of the body itself—by cutting some of it away or by adding something to it (in which event, it will not be the same body anyway). This immutable property of any body, this measure of its inertia, is called its **mass.** Rewriting equation (2) using a more felicitous notation, our Equation of Motion finally becomes:

$$F = ma$$

where m is the mass of the body.

(4) Units and Dimensions

Before algebra can be translated into arithmetic, we must have some agreement about how we are to measure our various quantities. Reverting to our constant-speed car of Fig. 22 (*a*), we wrote the equation:

$$x = Ct$$

But we found that the numerical value of C depended on whether we measured x in metres, or in kilometres. It would be also affected by the choice of seconds, or hours, for the units of t. In the present instance, the mass of a body may be evaluated from the fraction:

$$m = \frac{W}{g}$$

Depending on whether W is measured in tons, pounds, grams, kilogrammes, or anything else, and whether g is stated in units of length of feet, miles, metres or inches, and whether t is in seconds, minutes, hours or years, a whole range of different possible values of m will result. Putting it concisely, the value of m is determined by the units chosen for W and g.

In such a case, mass would be termed a 'derived' quantity (as distinct from a 'fundamental' quantity) since we have derived it from the force W. The gravitational acceleration g is also a derived quantity, being determined by the arbitrary choice of the units of length and time.

However, in a relationship of this kind it is quite arbitrary which quantity you elect to be fundamental and which becomes derived. In the system we are used to, length and time are considered to be fundamental, and velocity and acceleration are then derived from them. But 'fundamental' is a bad word here. There is nothing fundamental about length. For instance, the origin of the metre was a decimal fraction of the length of a quadrant of the Earth's surface, and the origin of the second was a fraction of the solar day. Both these quantities are quite arbitrary in choice and, at the time of their original selection, could not be measured with any great accuracy (or what we today regard as great accuracy). The important thing to appreciate is that, having chosen these two, you have *no choice for velocity or acceleration*, which *must* be expressed in the units of length and time already selected.

But it would be just as logical to adopt velocity and time as fundamental units. For instance, the standard velocity could be the velocity of light. Then the standard unit of length would become a derived unit, and would be the distance travelled by light in one second. This principle is actually embodied in the astronomical unit of distance called the light-year, which is the distance travelled by light in one year. By a similar argument, acceleration could be adopted as a fundamental quantity, along with time. The standard acceleration could be g. Velocity would then be derived from acceleration, and the standard unit could be the velocity attained by a freely falling body in one second. The unit of length could be the distance covered by a body travelling at this velocity in one second. Whatever the disadvantages of such a system, it might well have saved many generations of pupils and students many headaches.

For work in applied mechanics, there is a strong argument in favour of adopting mass as a fundamental unit. Mass is something tangible: you can hold a mass in your hand. Force is more elusive. You have to define your force in terms of 'the weight of a given mass' or 'the force to accelerate a certain mass a certain amount' or 'the force to compress a certain spring a certain distance'. So it is usually mass that is taken as a fundamental unit, while force is a derived unit defined in terms of a standard mass and Newton's second law.

If mass, length and time are chosen as arbitrary fundamental quantities, all other quantities in dynamics can be expressed in terms of these three. Mass, length and time are called the **dimensions** in which all other quantities are expressed. Thus, velocity can be expressed in terms of the dimensions of length and time. No matter what units are chosen, the equation

$$\text{velocity} = \frac{\text{length}}{\text{time}}$$

will always be true, because the units of velocity are defined by this relationship. If we adopt the foot as a unit of length and the second as the unit of time we shall always express velocity as feet/second, or feet per second. Acceleration, which is of course velocity/time, must then be expressed as feet per second per second.

If we now adopt the pound as our standard mass (by one pound I mean a piece of material having a weight on Earth of one pound), then our standard unit of force must be derived from the standard length and time, and equation (3). One unit of force must be the product of unit mass and unit acceleration. Writing this as an algebraic formula, and denoting 'unit' by a subscript u:

$$\begin{aligned} F_u &= m_u a_u \\ &= (\text{pound}) \times \frac{\text{foot}}{(\text{second})^2} \\ &= 1 \text{ pound foot second}^{-2} \end{aligned}$$

This may be easier to appreciate if we define unit force as 'that force which, applied to a mass of one pound, produces an acceleration of one foot per second per second'. Since we know that on Earth a mass acted upon by its own weight suffers an acceleration g, it follows that unit force, in any system, is $1/g$ times the weight of unit mass in that system. Using feet, seconds and pounds, unit force is 1/32·2 times the weight of one pound. Rather than call this 'one pound foot second^{-2}', we adopt a new name for it—the **poundal**. For calculations in dynamics, you must first distinguish clearly between the mass and weight of a body; secondly, you must express the mass and the weight in *units which are consistent with the Equation of Motion.*

All this has been complicated in the past by the unfortunate use of the same name for both the unit of mass and the unit of weight. I have already stated that one standard mass is 'one pound', which is commonly known as a unit of weight. What is more, people will continue to confuse the two terms. In common life, masses are evaluated by weighing them. To ordinary people, the distinction between two kilogrammes mass of apples and two kilogrammes weight of apples is academic, incomprehensible and unimportant. The only quick and practical way of evaluating a mass on Earth is by weighing it. One would hardly expect a greengrocer to apply a standard force to the apples and measure the resulting acceleration before he priced them. He very sensibly relies instead on a ready-to-hand force in the form of weight, which he compares with the weight of a standard mass. He would have to think more carefully, though, if his scale were a spring-balance type, and he were marketing apples on the Moon.

The net result is that no matter what system of calculation is to be used, the units of force (in which is included weight) are to be clearly distinguished from those of mass, and are also numerically different from them. The weight of one pound of mass is 32·2 poundals of force; the weight of one gramme of mass is 981 dynes of force. For engineers, who find the pound force too familiar to reject, a new standard of mass has been devised: it is called the **slug**, and is a mass having a weight of approximately 32·2 pounds force. The following table gives four of the more familiar systems of units in use.

System	*Mass*	*Length*	*Time*	*Force*	*g*	*Remarks*
Absolute	pound	foot	second	poundal	32·2 ft/s^2	unit of force, approx ½ ounce
Engineer's	slug	foot	second	pound	32·2 ft/s^2	unit of mass, approx 32·2 lb
c.g.s.	gramme	centimetre	second	dyne	981 cm/s^2	—
MKS	kilogramme	metre	second	newton	9·81 m/s^2	now being widely adopted

There are many arguments to support the adoption of the M.K.S. system. The units of force and mass are both of a reasonable magnitude. Unit mass (1 kg) is a little over 2 pounds, and unit force (1 newton) is something under ¼ pound. Moreover, the system is particularly useful in electrical engineering, since it yields the practical units of volt, ampere etc. It also conforms to the process of metrication, and most textbooks in mechanics and engineering subjects now adopt this system. We have already used the newton as a unit of force, and now, rather belatedly, we are able to define it as that force which, when applied to a mass of one kilogramme, causes an acceleration of one metre per second per second. Since g is 9·81 metres per second per second, it follows that one newton is 1/9·81 of the weight of one kilogramme, or 0·224 pounds in English units.

(5) Some Applications of the Equation of Motion

I have stressed that this book is not intended as an instruction manual in applied mechanics, but rather as an attempt to show how the engineer makes use of applied mechanics in his work. One of the difficulties is that, in order to do this, I am obliged to use illustrations which might at first sight appear pointless and trivial. However, if you set out to learn a new language (which is really what you have to do when studying this subject) you cannot expect to write impressive and glowing prose after only a few lessons. You have to subsist on simple words and sentences. It is not my intention to teach you how to calculate the power of the driving turbines of an ocean liner, or the fuel system of a rocket. But I hope that you will at least learn how the marine engineer, and the space engineer, set about determining these things for themselves.

As a first example, let us imagine a cannon pointing vertically upwards. The barrel has a length of 6 metres, and the cannon fires a shell which has a mass of 200 kilogrammes. We shall assume for simplicity that the shell is driven along the cannon's barrel by a propelling force which is constant for the whole passage of the barrel, and has a value of 3 Meganewtons (i.e. newton $\times$ 10^6). This assumption of a constant force is grossly incorrect. The force is bound to decrease considerably as the gas in the barrel expands as the shell is driven along. Let us find how high the shell will rise before it begins to fall back to earth, and for further gross simplification, we shall assume that it has no air resistance to contend with.

This problem must be solved in two stages. While in the barrel of the gun, the shell is subjected to one system of forces. After leaving the barrel, it becomes subject to another system. Each phase must have its own equation of motion, and each equation of motion can be obtained most easily by drawing the free-body diagram for the shell. While it is in the barrel, the shell is subjected to two different forces: the propelling force, upwards, and the weight, downwards. This results in an acceleration a_1. After leaving the barrel, the only force acting is the downward weight, and this results in a second acceleration a_2. The two diagrams are shown in Fig. 24.

Although we know the actual values, I have designated the propelling force by P and the weight by W. The reason is that most problems of this kind can be solved more easily and with less likelihood of error if the calculation is performed algebraically, leaving the arithmetical values to be substituted at the end.

For Fig. 24 (a) the resultant force is clearly upward, and the equation of motion is:

$$P - W = ma_1$$

Remember that F, in the general equation, is the net or resultant force on the body. P is given. W is obtained from the product of m and g. As usual, the value of g may be taken as 9·81 metres per second per second. However, substituting mg for W, and rearranging:

$$P - mg = ma_1$$

Therefore $$a_1 = \frac{P}{m} - g$$

$$= \frac{3 \times 10^6}{200} - 9{\cdot}81$$

$$= 15\,000 - 9{\cdot}81$$

$$\simeq 15\,000 \text{ metres per second per second}$$

The value of 9·81 is so small that we can neglect it in this calculation. In physical terms, the weight of the shell is so small in comparison with the

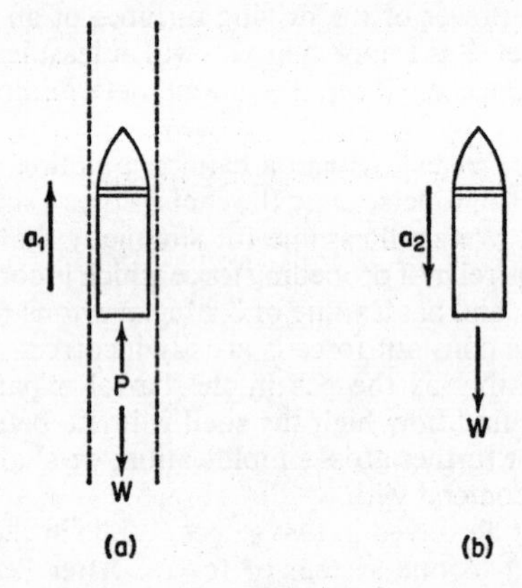

Fig. 24. Free-body diagrams for (*a*) the shell in the barrel, and (*b*) the shell in flight.

propelling force that it may be neglected *for the purpose of determining the muzzle velocity*. But bear in mind that we cannot afford to neglect the weight after the shell has left the barrel, when it becomes the *only* force acting.

The shell starts from rest at the bottom of the barrel, and suffers this acceleration for a distance of 6 metres. We can now make use of one of the equations of kinematics stated in Chapter Two. The relevant equation is:

$$v^2 = u^2 + 2ax$$

v is the final (muzzle) velocity we require; u is the starting velocity (zero in this example); x is the distance (6 metres); a is the acceleration we have just found. Substituting values:

$$v^2 = 0 + 2 \times 15\,000 \times 6$$

$$= 180\,000$$

Hence $$v = 425 \text{ metres per second}$$

Now we turn to Fig. 24 (*b*). Since there is only the one force, the resultant acceleration a_2 must be in the direction of this force, and is so shown. The new equation of motion is:

$$W = ma_2$$

and substituting as before:

$$mg = ma_2$$

from which a_2 is clearly g. (In fact we already knew that $a_2 = g$, because the shell is in a state of free 'fall' even though it is travelling upwards initially.) We can again use the equation:

$$v^2 = u^2 + 2ax$$

In this case, v is zero (as the shell comes to rest instantaneously at the top of its flight); u is the initial velocity, which we have just worked out (as v for the first phase of motion); a is $a_2 = g$; and x is the height we wish to find. But we must pay due attention to the signs of our quantities. If we adopt a sign convention 'upwards is positive', we must remember to call our acceleration negative (because it is in the downward direction). Substituting:

$$0 = u^2 + 2 \times (-g) \times x$$

$$x = \frac{u^2}{2g}$$

$$= \frac{(425)^2}{2 \times 9{\cdot}81}$$

$$= 9170 \text{ metres}$$

Consider next the very simple hoist shown in Fig. 25. This consists of a loaded cage of total mass 1000 kilogrammes, and a counterweight of mass 800 kilogrammes. The two are connected by a rope, which we shall assume has no mass at all. We shall also make the following assumptions: that the loaded cage is to be raised from rest; that for the first part of the lift, it requires an upward acceleration of 2 metres per second per second; that for the final part of the lift (as it slows down to rest at the top), it requires a downward acceleration of the same value. Note in passing that the acceleration again may be downwards even though the velocity is upwards, as it was for the shell. What net force will be required at the loaded cage, to produce each of these accelerations?

This problem differs from the last in that we now have two bodies: the cage and the counterweight. We have to draw a free-body diagram for each one. The cage is subjected to three forces: the weight (downwards) called W_A; the tension in the support rope, which must act upwards, and is called T_1 and T_2 for the two stages of the motion; and the propelling force, which I have assumed upwards, and have called P_1 and P_2 for the two stages. These forces are shown on the free-body diagrams for the two bodies, and for the two conditions of acceleration. Because we know the acceleration in this example, this also is shown.

For the upward acceleration, we obtain the following two equations:
for the cage

$$P_1 + T_1 - W_A = m_A a$$

for the counterweight

$$W_B - T_1 = m_B a$$

Notice that the direction of the resultant force is in the direction of the acceleration, and that the acceleration is equal in magnitude for both bodies but opposite in direction. The above are simultaneous equations, and we can

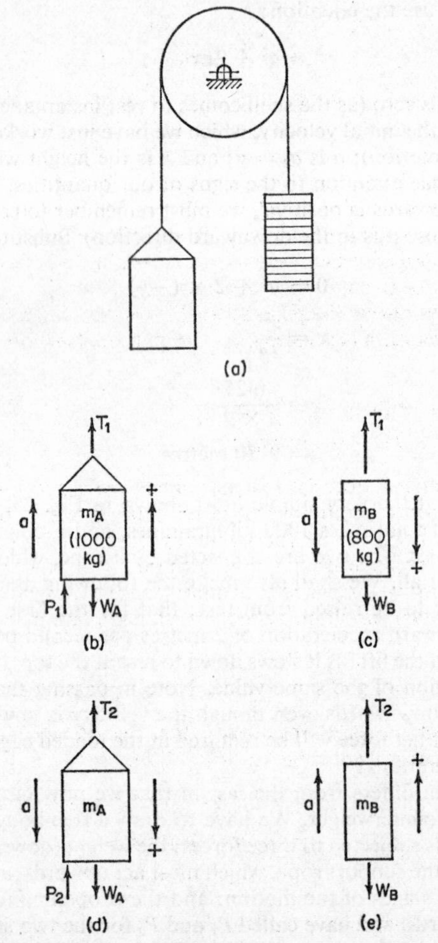

Fig. 25. A simple hoist with counterweight: (*b*) and (*c*) show the free-body diagrams for upward acceleration of the cage. (d) and (e) show the diagrams for downward acceleration of the cage.

solve them by eliminating T_1 in the usual way, pausing to note that the tension T_1 must *not* be assumed to be the same as the weight of the cage. This gives us:

$$P_1 = m_A a + m_B a + W_A - W_B$$

But we know that $W_A = m_A g$ and $W_B = m_B g$. Substituting these, our equation then becomes:

$$P_1 = a(m_A + m_B) + g(m_A - m_B)$$

Substituting known values:

$$\begin{aligned} P_1 &= 2(1000 + 800) + 9{\cdot}81(1000 - 800) \\ &= 3600 + 1962 \\ &= 5562 \text{ newtons} \end{aligned}$$

I have deliberately solved this problem by a roundabout method, because this is the way it would have to be solved if we were given the force P_1 but not the acceleration. Actually, the second of our equations of motion may be solved by itself. If you try this, you should obtain a value for T_1 of 6248 newtons. The weight of m_B is $800 \times 9{\cdot}81 = 7848$ newtons. So we see that the tension T_1 in the rope is *not* equal to the weight, being in this case less. If the tension upwards were exactly equal to the weight downwards, there would be no resultant force to cause the acceleration.

We may now repeat the process for the second stage of the motion. This time, the acceleration is reversed, resulting in the two following equations of motion:

$$W_A - P_2 - T_2 = m_A a$$

$$T_2 - W_B = m_B a$$

It should be noticed, both for this pair of equations and for the first pair, that forces are reckoned positive when in the same direction as the acceleration. This direction is shown on the free-body diagrams by an arrow with a + sign. But this positive direction is arbitrary. We could choose the other direction as positive, provided the equations were altered to suit.

Solving our two simultaneous equations gives us the following expression for P_2:

$$P_2 = -a(m_A + m_B) + g(m_A - m_B)$$

which is almost the same result as before, except for the negative sign in front of a. Again substituting known values:

$$\begin{aligned} P_2 &= -2(1000 + 800) + 9{\cdot}81(1000 - 800) \\ &= -3600 + 1962 \\ &= -1638 \text{ newtons} \end{aligned}$$

This negative answer means that the force P_2 must be applied in the opposite direction to the assumed one; in this case, downwards instead of upwards. If no force P were applied, there would still be a downward acceleration, but it would be less than 2 metres per second per second. This serves to highlight an important point in the handling of a mine cage—or any other form of hoist. It might be cheerfully assumed that all you have to do is to apply an upward force to the cage, in order to lift it, and that its own weight will stop it at the top. But if you apply the upward force for too long, it may well be travelling too fast for its own weight to bring it to rest at the top; the consequence need not be described.

Finally, let us consider a modern jet aircraft ascending at an elevation of 30°, reaching a speed of 120 metres per second, from rest, in 30 seconds. The mass of the plane is 5 tonnes (5000 kg), and average resistance of the air to its motion is 20 kilonewtons. What engine thrust is required for this performance?

The free-body diagram in this case discloses four forces, as shown in Fig. 26. The engine thrust T obviously pushes the plane forwards. The weight W acts vertically downwards. The air resistance or drag, denoted by D, is along

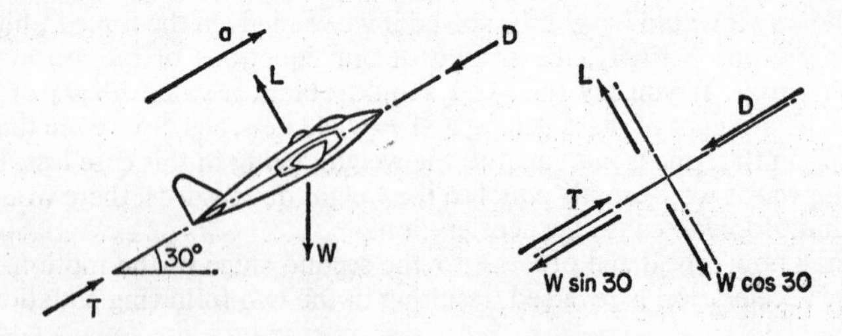

Fig. 26. The free-body diagram of an aircraft in flight.

the direction of flight, but reversed. The motion through the air also causes lift, an upward thrust on the wings and tailplane, shown as L on the diagram. As specified, the acceleration a is along the direction of flight.

The equation of motion must always be along the direction of the acceleration, and so we require the resultant force along this direction. The method of resolving forces is now to be used. Although the weight W does not act along the flight path, it has a component along this line. We therefore resolve W along two directions: one along the path of flight, and the other at right-angles to it. The system of forces, with this process completed, is also shown in Fig. 26. We are now able to write the equation of motion as

$$T - W \sin 30 - D = ma$$

from which we get

$$T = ma + D + W \sin 30$$

The mass m is given. The acceleration a is not given, but we can work it out from the data. Using another of the useful equations of kinematics from Chapter Two:

$$v = u + at$$
$$120 = 0 + a \times 30$$

giving a value of 4 metres per second per second for a. The air resistance D is given. Finally, $W = mg = 5 \times 10^3 \times 9{\cdot}81$ newtons. Substituting these values in our equation of motion:

$$
\begin{aligned}
T &= (5 \times 10^3)4 + (20 \times 10^3) + (5 \times 10^3 \times 9{\cdot}81)\,0{\cdot}5 \\
&= 10^3(20 + 20 + 24{\cdot}5) \\
&= 64{\cdot}5 \text{ kilonewtons (or approximately 14 500 pounds force)}
\end{aligned}
$$

Notice how all the decimal multiples of newtons and kilogrammes are converted before the calculation is made, the final conversion back to kilonewtons being a matter merely of numerical convenience. Notice also that we never needed to write an equation in the direction at right-angles to the flight path to determine the engine thrust. An equation could be written here, but since no acceleration takes place along this direction, it would be an equation of equilibrium and not of motion.

(6) Centrifugal and Centripetal Force

We saw in Chapter Two that uniform motion in a circular path results in an acceleration directed towards the centre. Applying Newton's Second Law to this situation, we find that a body moving in such a path must be subjected to a force which acts towards the centre of rotation. This force is called the **centripetal force**. It must be provided by some sort of constraint on the body which compels it to follow the circular route instead of flying off at a tangent along a straight line. An ordinary car travelling round a bend relies on the friction between tyres and ground to keep it on the road. A racing driver requires to go round at a much greater speed and hence greater acceleration. Friction is usually insufficient, and so the track may be banked. The reaction of the track on the car is then not vertical, but inclined. It has a vertical component which balances the weight of the car, and also a horizontal component which provides the required inward thrust. A locomotive relies on the pressure of the outer rail on the flange of the wheel, and also on slight banking of the track (called super-elevation). A planet revolving around a sun relies on the pull of solar gravity.

If you have ever been given to understand that a body moving in a circular path is subjected to a radially outward force called the centrifugal force, I urge you most strongly to forget it and to start re-thinking your ideas of force. We have shown clearly that the motion consists of a centrally directed acceleration; therefore, if we are to believe Newton, the *resultant force on the body must also be centrally directed*. It is at this stage that I must reassert the warning I gave in Chapter Three concerning forces. Forces are real things and cannot be willed into existence because of the existence of a certain type of motion. The shell fired from the cannon in our example of Section 5 is in high-speed upward flight originally, but the only force acting on it after it leaves the barrel is the downward-acting weight. Force cannot be brought to bear on a body merely because of its motion in a circle. This is putting the cart before the horse. Force is the *cause* of circular motion—not an effect caused *by* it. As the Earth spins round the Sun, the one and only relevant force acting on the Earth is the pull of solar gravity. There is certainly no centrifugal force pulling the Earth the other way.

Then why, you may ask when travelling in a coach round a bend, are you thrown outwards? What force is it that pushes you outwards? The rather unfair answer is that you are asking the wrong question: what you should ask is 'Why don't I move round in a circle with the coach?' Then, I hope, the answer is clear: that there is no force pushing you round. If you stand against the outer wall of the coach, you will go round with it. But then the outer wall of the coach is pushing you inwards, not pulling you outwards. Centripetal force is a real force: it is a force which is always directed towards the centre of the path, and it must be accounted for by real tangible things. If it cannot be so accounted for, the circular motion cannot exist.

So are we to assume that centrifugal force is a mere myth—a figment of the imagination, designed to puzzle students? No, it is not. Centrifugal force is also a real force, and it *does* act *outwards from* the centre. But it *does not act upon the body*. It is, as predicted by Newton's Third Law, the equal and opposite reaction exerted *by* the body *upon* its counterpart in the system. The frictional force between tyres and road pushes the car *towards* the centre. But the same frictional force pushes the road outwards away from the centre. (Notice which way the gravel flies when a car takes a bend at high speed.) The outward thrust of the racing driver's wheels is the centrifugal force applied *by* the car *upon* the road. On a railway track, the rails are pushed outwards by the centrifugal force exerted *by* the locomotive *upon* them. The solar planet exerts an outward centrifugal force *upon* the sun.

To attempt to draw the free-body diagram of anything moving in a circular path, inserting the so-called centrifugal force as an outwardly directed force on the body, is a pernicious practice which serves always to confuse and never to clarify. The fact that the 'right answer' can be obtained by this practice is no justification for its use in teaching dynamics. The right answer is obtained, first by popping in this non-existent force, and then by writing an equation not of motion, but of equilibrium. This is an example of the use of what is called D'Alembert's principle. It is a perfectly admissible procedure if you are sure what you are doing, but it is a dangerous practice to allow a novice to use. Any skilled practitioner of any art or science is entitled to use tricks and short cuts in his work, but he should be wary of passing them on to his pupils. The result may well be, and indeed often has been, bad habits of thought and understanding which take years to overcome.

Calculations involving circular motion are carried out in the same manner as our three earlier examples. A free-body diagram is drawn and the equation of motion written, positive in the direction of the acceleration. We know that the acceleration is towards the centre, and we found in Chapter Two that it has a magnitude of v^2/R. Imagine, for example, a car travelling round a circular track of mean radius 10 metres. The sideways frictional force between tyres and road may be assumed to have a maximum possible value of 0·4 times the weight of the car. At what maximum speed may it negotiate the track without skidding sideways?

The free-body diagram in Fig. 27 shows only three forces: the weight W, acting downwards; the upward reaction U of the ground on the wheels; and the frictional F, which acts from right to left, assuming that the centre of the track is to the left. We know that the acceleration in this case is the centripetal acceleration and has a magnitude of v^2/R. There may also be an acceleration in the direction of travel of the car, but this does not affect our calculation.

The equation of motion is thus:

$$F = ma = m \times \frac{v^2}{R}$$

But $\quad F = 0{\cdot}4W$ and $m = \dfrac{W}{g}$ so that:

$$0{\cdot}4W = \frac{W}{g} \times \frac{v^2}{R}$$

Therefore

$$v^2 = 0{\cdot}4gR$$
$$= 0{\cdot}4 \times 9{\cdot}81 \times 10$$
$$= 39{\cdot}24$$

Hence

$$v = 6{\cdot}26 \text{ metres per second}$$

If the assumed data are correct, the vehicle will not be able to negotiate the track if driven at a higher speed than this, because the frictional force inwards is insufficient to provide the acceleration. It is interesting to observe

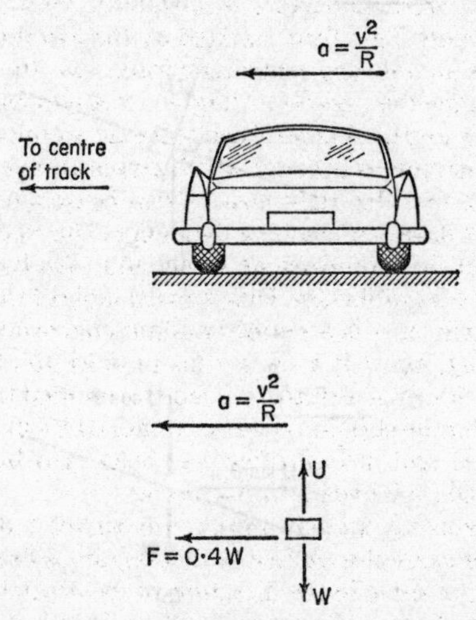

Fig. 27. Free-body diagram of a car travelling round a circular track.

that the limiting speed was obtained without specifying the mass of the car. The reason is that the maximum frictional force is usually a fraction of the weight, and hence is proportional to the mass.

Suppose now that a racing car is required to travel at a steady 30 metres per second around a circular track of mean radius 100 metres. Let us calculate the angle at which we should have to bank the track in order not to rely on friction at all. The diagram is shown in Fig. 28.

Since we must assume no friction, there are only two forces acting. The direction of the acceleration is horizontally towards the centre of the track (to the left). Before writing the equation of motion, we have to resolve the reaction force U into components along this line and at right-angles to it. The resolved force system is shown in Fig. 28 (*c*). We now write the equation of motion:

$$U \sin \theta = ma = m \times \frac{v^2}{R}$$

But we cannot evaluate θ from this single equation, because we do not know U or m. In such a case, we have to write a second equation along the other

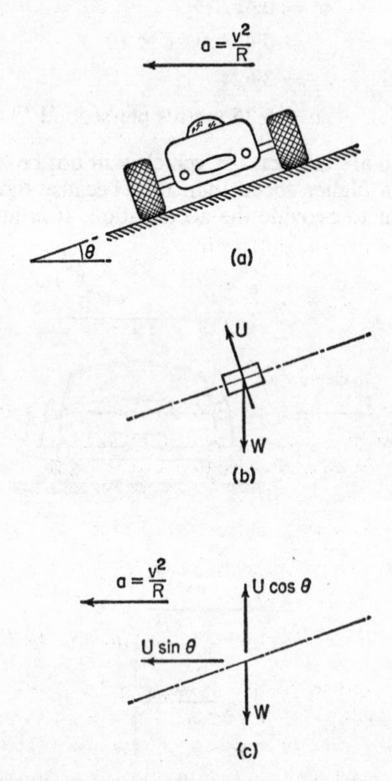

Fig. 28. Free-body diagram of a car on a banked track with no friction.

direction. You can think of this as a second equation of motion, if you like, but the acceleration along the vertical is zero anyway, and so the equation of motion becomes an equation of equilibrium. The equation is:

$$U \cos \theta = W$$

Dividing the first equation by the second, in order to eliminate U, we get

$$\frac{\sin \theta}{\cos \theta} = \tan \theta = m \times \frac{v^2}{R} \times \frac{1}{W}$$

But since $W = mg$ we can substitute W/g for m. Therefore

$$\tan\theta = \frac{W}{g} \times \frac{v^2}{R} \times \frac{1}{W}$$

$$= \frac{v^2}{Rg}$$

Substituting the given values:

$$\tan\theta = \frac{(30)^2}{100 \times 9{\cdot}81}$$

$$= 0{\cdot}917$$

which gives a value for θ of $42\frac{1}{2}°$.

Let us turn finally to a rather topical problem, that of a satellite orbiting the Earth. Clearly, the only operative force in this example is the earth-weight of the body. (This is neglecting any resistance due to remnants of air, and also any gravity effects due to other heavenly bodies.) If we call the mass m, the weight is mg, and it is this force that must cause the centripetal acceleration. The equation of motion is therefore:

$$mg = m \times \frac{v^2}{R}$$

from which

$$v^2 = Rg$$

If the Earth's radius is assumed to be 6500 kilometres and we neglect the 50 or 60 kilometres of extra radius due to the height of the flight path above the ground, we can arrive at an approximate value for v, the least velocity to keep the satellite in orbit.

$$v^2 = Rg$$

$$= (6500 \times 10^3) \times 9{\cdot}81$$

$$= 63{\cdot}2 \times 10^6$$

Therefore $v = 7950$ metres per second (or 17 800 miles per hour)

This is the minimum velocity for orbiting the Earth, and a simple calculation shows that the time required for one orbit is about 85 minutes. Again, we have had to make certain approximations for the sake of simplicity. For instance, g is taken as 9·81 metres per second per second, but the value would be less than this at the altitude usually attained by a satellite. Also, we have assumed a circular path, whereas the true path is an ellipse. But our answer is approximately correct, and you can appreciate that before a body can go into orbit it must attain this minimum velocity; if it fails to do so it will fall back again to the ground. If it is required to travel away from the Earth altogether, it must attain a slightly higher velocity, called the escape velocity.

While it is orbiting the Earth, the satellite or spacecraft is in a state of free 'fall', so to speak. In other words, the only force acting on it is its weight, and it is not resisting this weight but is being accelerated by it in exactly the same manner as a body which is dropped freely from a height. Exactly the same conditions apply to any object or person within the satellite. The *only* force acting on a member of the crew is his earth-weight (or something slightly less, because of his altitude), and it is this force which is causing his orbital motion. Contrast this state with the man standing on the ground, when the downward force of his weight is balanced by an equal upwards reaction from the ground. There can be no reaction between the astronaut and his craft, which is moving under exactly the same conditions. His feet will not press on the floor of the craft, any more than your feet would press on the floor of a lift if you happened to be in one that was falling freely down the shaft. In the case of the lift, the situation would only apply until the lift reached the bottom of the shaft; but the satellite continues in a steady state of orbital motion, and its crew are said to be in a state of weightlessness. This is an unfortunately misleading term, because their weight is the only force they are experiencing. This weight is practically the same as their earth-weight but, because they are in a state of free 'fall' *around* the Earth, the sensation exactly corresponds to that they would experience if they were in an environment removed from all gravity.

(7) Rotational Motion and Moment of Inertia

In examining the kinematics of motion in Chapter Two, we mentioned the case of spinning bodies—bodies which move, but do not change position in space. Examples include electric motors, turbines, car engines, and ship's propellers; all of which start from rest and attain high rotational speeds independent of any bodily movement they may have. Because these bodies have mass, force will be needed to cause their rotational acceleration. But we cannot apply the equation of motion in its ordinary form, $F = ma$, because this equation is the algebraic equivalent of the results of experiments on bodies moving in a straight line. We have to deduce an alternative form of the equation for the case of rotating bodies.

We will start by considering the very simple rotating body shown in Fig. 29 (*a*). It consists of a single mass m at the end of a light rod which is in turn attached to a shaft. We shall cause this shaft to accelerate by applying a torque, or twist, to it. Remember that the acceleration is an angular one, in radians per second per second; we shall call it α, the Greek letter alpha. The magnitude of the torque is T, and the radius of the circular path of the mass is r.

We showed earlier that the tangential linear acceleration of the mass m is related to the angular acceleration by the equation:

$$a = \alpha r$$

So the mass m must be subjected to a force which we can obtain from the Equation of Motion:

$$F = ma = m \times \alpha r$$

This force is applied to the mass, via the rod, from the torque on the shaft. So the torque T applied to the rod must be the product of the force F and the radius r. Hence

$$T = (m\alpha r)r = \alpha(mr^2)$$

Now we shall complicate the case by mounting several masses, m_1, m_2, m_3 etc. at various respective radii r_1, r_2, r_3, as shown in Fig. 29 (*b*). We can see that, in order to accelerate this shaft, each separate mass will require its own torque, which can be calculated from the formula above. For such a system of multiple masses, the required torque would be:

$$T = \alpha m_1 r_1^2 + \alpha m_2 r_2^2 + \alpha m_3 r_3^2$$

and so on for as many masses as we have.

The torque T can best be thought of as a kind of rotational force, and α is a rotational acceleration. So the equation we have derived is a sort of rotational equation of motion, in which the force F is replaced by the torque T, the linear acceleration a is replaced by the angular acceleration α, and the

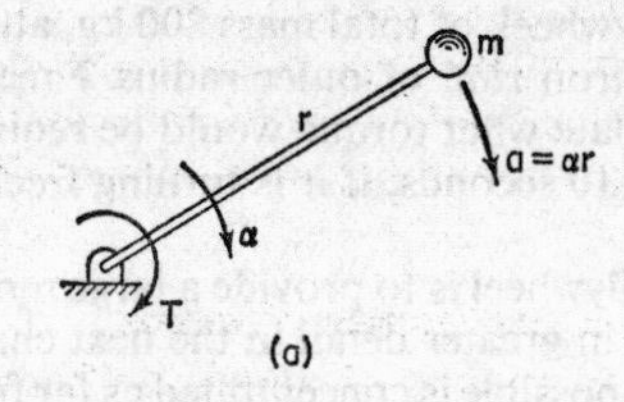

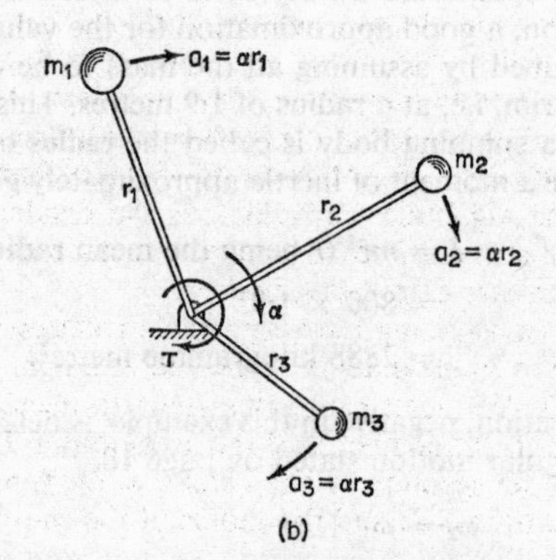

Fig. 29. (*a*) Single rotating mass. (*b*) System of rotating masses.

mass m is replaced by ($m_1 r_1^2 + m_2 r_2^2 + m_3 r_3^2$ etc.). This last expression is a rather untidy mixture of terms, and it represents some quantity which is dependent upon the *way the total mass is distributed about the centre of rotation*. It is called the **moment of inertia**, denoted by I.

There are various ways of finding the moment of inertia for any rotating body, provided that its shape and its mass (and hence its mass distribution) are known. One is the method we have used above, but unfortunately few spinning bodies group themselves into discrete masses spinning at their own individual radii. Most spinning bodies have their mass distributed continuously over a range of radii. If we think, for instance, of a solid uniform disc, the mass is distributed all over, from a radius of zero to the outer radius of the disc. For a uniform shape such as a disc, we use the calculus to evaluate the moment of inertia. The value is $\frac{1}{2}mR^2$, where R is the disc's outer radius.

Textbooks and handbooks of mechanics usually contain tables giving the values of moments of inertia for standard bodies of regular shape. With bodies that are not quite regular in shape, such as a car wheel, a calculation will give an approximate value; but the best way of obtaining I is by experiment, e.g. applying a torque and measuring the resultant acceleration.

The moment of inertia of a rotating body may be thought of as the rotational equivalent of its mass, in the same way that the torque applied to it can be considered as the rotational equivalent of force. The equation of motion for rotation is thus:

$$T = I\alpha$$

Imagine a large flywheel, of total mass 800 kg, attached to a shaft. Suppose it has a heavy cast-iron rim, of outer radius 2 metres and inner radius 1·8 metres. Let us calculate what torque would be required on the shaft to bring this wheel to rest in 10 seconds, if it is turning freely at 1200 revolutions per minute.

The purpose of a flywheel is to provide a large moment of inertia. (We shall examine this matter in greater detail in the next chapter.) For this reason, as much of the mass as possible is concentrated as far from the centre as possible, i.e. in the rim. The spokes are merely there to attach the heavy rim to the shaft. For this reason, a good approximation for the value of the moment of inertia will be obtained by assuming all the mass to be concentrated at the mean radius of the rim, i.e. at a radius of 1·9 metres. This mean, or effective, dynamic radius of a spinning body is called the **radius of gyration.** The flywheel thus will have a moment of inertia approximately given by:

$$I = m\bar{r}^2 \text{ (}\bar{r}\text{ being the mean radius)}$$

$$= 800 \times 1{\cdot}9^2$$

$$= 2888 \text{ kilogramme metre}^2$$

The angular acceleration, negative in this example, is determined from one of the formulae of angular motion stated on page 18.

$$\omega_2 = \omega_1 + \alpha t$$

$$0 = 2\pi\frac{1200}{60} - \alpha \times 10$$

Hence $\alpha = 12{\cdot}56$ radians per second per second

(Remember that the angular velocity has to be converted to radians per second.)

Using the rotational equation of motion:

$$T = I\alpha$$

$$= 2888 \times 12{\cdot}56$$

$$= 36\,200 \text{ newton metres}$$

How this torque is to be applied is outside the province of the question. If the wheel were brought to rest by a tangential force at the rim, for instance, the torque would be the product of the tangential force and the rim radius.

If we call the force F:

$$F \times 2 = 36\,200$$

Hence $$F = 18\,100 \text{ newtons}$$

On the other hand, the braking force may be applied by a brake-shoe on a drum attached to the wheel shaft. If the drum radius is 10 centimetres:

$$F \times 0{\cdot}10 = 36\,200$$

$$F = 362 \text{ kilonewtons}$$

CHAPTER FIVE

WORK, ENERGY, POWER AND MOMENTUM

We dealt in the previous chapter with what we might call the birth of modern dynamics, which started with the formulation of Newton's three laws of motion and their application to practical problems. But it must not be assumed that the whole of dynamics rests exclusively on this basis. The examples of calculations were concerned either with evaluating a force to produce a specific motion, or with predicting the motion resulting from a known system of forces. But we can propound many examples in which the Equation of Motion could either not be formulated, or, if it could, would not lead us to a solution.

For instance, we examined the problem of calculating the required thrust of a jet engine. Suppose we had asked the further question: 'At what rate must the exhaust gases be propelled backwards to produce this thrust?' Or again: 'How much fuel would an Atlantic liner require to make the crossing from England to America? And would it require more if it made the crossing faster? And if so, why?' For the answer to these, and many other problems, we have to examine some further principles of dynamics. An answer to the first question is to be found in the principle of momentum, which we shall shortly see is derived from a consideration of the Second Law of Motion. The problem of the liner requires a consideration of power and energy. Newton seems to have had little to do with energy, leaving the main principles to be worked out much later on by Carnot, Clerk-Maxwell and others.

(1) Energy and Work

A good starting point for the understanding of energy is the popular meaning of the word. When we describe a youngster as 'bursting with energy' we know that he is anxious to do things—active things, not reading, or anything requiring sitting still. He wants to run about, climb trees, play football, or cricket. There is something inside him that is clamouring for release, and which can be released only by physical activity.

And so it is with mechanical energy. Its correct definition is 'the capacity to do work', but we must limit the meaning of the word 'work' to its mechanical sense. In mechanics, work is only done when bodies move, and furthermore, the movement must be exerting a force. A student may work hard, sitting at a desk and reading for an examination; but in the mechanical sense of the word, he is doing no work at all. He may tire of reading and go out for a walk. If his walk takes him along a level road, he is again doing practically no work, in the mechanical sense. If he goes down a hill he is still doing no work; in fact, he is having the work done on himself. Only if he elects to climb a hill does he have to exert a force in the direction of his movement, and so do mechanical work. He exerts a forward force to overcome the backward component of his weight.

As always, we have to define our terms quantitatively, so that we can measure them. For work to be done, two things are needed: a force and a movement. Our definition of work is thus the product of the force and the movement. Calling the force F, the movement x, and the work J, we can write:

$$J = Fx$$

We have chosen to measure F and x in newtons and metres, respectively, and so our unit of work is a derived unit. We could call this the newton metre but we prefer to call it the **joule**, after one of the pioneers of the principle of energy in thermodynamics.

To sum up, then, we can write a specific definition of work as follows: 'Work is said to be done when a force moves its point of application in the direction of the said force. Work is the product of the force and the distance moved.'

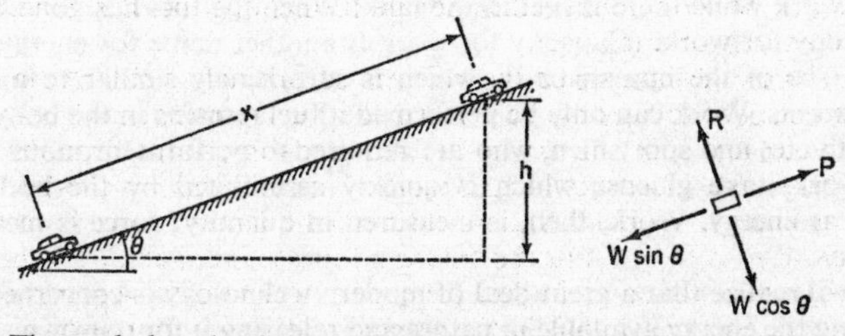

Fig. 30. The work done by the car ascending the hill is either the product of weight and vertical height, or the product of tractive force and distance x. Both are the same, if friction is excluded.

In Fig. 30 a car is ascending a hill. The vertical height of the hill is h and the inclined distance travelled by the car is x. We shall assume for the sake of simplicity that the only forces acting are the weight W of the car, the tractive force P exerted by the engine, and the reaction R of the road upon the car. We shall neglect any frictional force at the wheels, or any resistance due to the motion of the car through the air. The free-body diagram on the right of Fig. 30 shows the force system, with the weight W resolved into components parallel to, and perpendicular to, the line of motion. We shall also assume that the car ascends the hill at uniform speed, so that the acceleration is zero. We can now equate forces in each of the two directions, thus:

$$P = W \sin \theta$$

$$R = W \cos \theta$$

where θ is the angle of the incline.

The work done by the car engine is the product of the force P and the distance moved x. Hence:

$$J = Px$$

$$= (W \sin \theta)x$$

But you can see from the triangle that $\sin \theta = h/x$

$$J = W\frac{h}{x}x = Wh$$

so that the work done is exactly the same as if the car had been lifted vertically upwards by a force equal to its weight. This result may already give an inkling of the advantages of an energy approach to a problem. The notion of work tells us that the work done in raising the car from the bottom to the top of the hill is *unaffected by the way in which this is done.* We actually took the car up an incline, but the same work would have been done if the road had been a winding track with variable slopes, or even if it had started off going downhill. The work done is only determined by the difference between the starting point and the finishing point. I shall return to this aspect later, on page 71.

The vital point to appreciate about work is that there is a definite limit to it, according to the situation. There is no theoretical limit to force. With suitable gearing, a car can exert any force you care to specify. But the car can only do work while there is fuel in the tank. When the fuel has gone, so has the capacity for work. (Capacity for work is another name for energy.) The same is true of the human body, which is surprisingly similar to a car in many respects. Work can only be performed if fuel remains in the body. This is why athletes and sportsmen, who are required to perform arduous mechanical work, take glucose which is quickly assimilated by the body and released as energy. Work, then, is measured in quantity; force is measured in degree.

You will realize that a great deal of modern technology is concerned with harnessing the energy available in nature and releasing it for our own specific purposes. The most obvious energy sources that spring to mind are coal and petroleum. Both are sources of chemical energy. Before this can be made to perform work, it has to be converted to heat, which is another form of energy. A relative newcomer to the scene is atomic energy, and here also we use the intermediate stage of heat. In a petrol engine, the conversion takes place in the cylinder where the fuel is burned. The resulting heat expands the gases, causing pressure, which drives the piston. The same action takes place in a diesel engine, though the fuel is oil. A power station converts the chemical energy of coal or fuel oil into heat, which is then used to convert water into high-pressure steam, which in turn is caused to perform mechanical work in a turbine. We shall see later that a gas or a liquid under pressure is a source of energy. Also, a spring or a deformed steel bar is capable of releasing energy, called strain energy. Dynamics, however, is concerned only with the motion of *rigid* bodies, so energy is manifested in only two forms: potential energy and kinetic energy.

(2) Potential Energy and Kinetic Energy

Potential energy is the energy a body gains due to its being raised up from one level to another. Suppose, for instance, you hoist a heavy cage from the bottom of a mine shaft to the top. You must perform work to do this, and so you have given the cage some energy. Because you have given the cage some energy, it is now in a position (at the top of the shaft) where it could perform some work itself. For example, you could attach it by a rope to an identical cage at the bottom of the shaft and, without doing any further work, raise the lower cage to the top merely by letting the first cage discharge the energy it has accumulated. This is actually done, in effect, in one type of hoist: two identical cages are hung from the ends of a single rope which passes over a pulley at the top, so that as one cage ascends the other descends. If the cages are of equal weight, the system will, theoretically at least, require no energy to operate it; in practice, of course, work must be done to overcome friction. An ordinary lift cage has a counterbalance weight that is roughly the same weight as the cage, so that the work performed in raising the empty cage is practically negligible. The real work done by the driving motor is in raising the contents of the cage, not the cage itself.

It is a fundamental principle of energy that you cannot get rid of it or produce it from nothing. This is known as the principle of **conservation of energy**. It has no proof, but is based on the accumulated experience of centuries. So if you do work on a body, the body must gain energy equal to the amount of work that you do on it. This fact enables us to calculate the potential energy of a body. All we have to do is to evaluate the work required to raise the body through the relevant height.

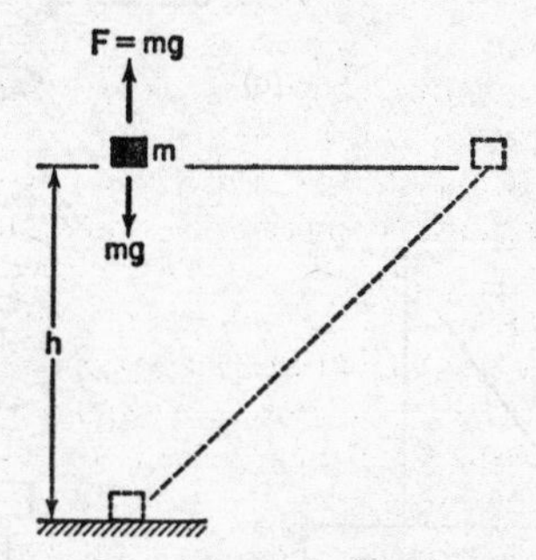

Fig. 31. The potential energy of the mass m is the work required to raise it through the height h.

In Fig. 31 we have caused the mass m to be raised through a vertical height h above the lower point. In doing so, we must apply a vertical force to overcome the downward-acting weight mg. So the work done is the product $mg \times h$ and this is the potential energy:

$$J_{pot} = mgh$$

If we had raised the mass along the dotted path instead of directly vertically, the work done would have been the same—as we have already found out with the car in Fig. 30. The question now arises, what is the height h which defines the potential energy of a body? The answer is that it is determined by circumstances.

In Fig. 32 (a) water in the lake on the mountain top has potential energy which may be employed to generate electricity. In such a case, the height h used to evaluate the energy is the distance between the level in the lake and the power station in the valley below, because this is as far as the water may fall to be of use to the engineer. It may fall further, on its subsequent journey to the sea, but this is of no interest to the designer of the power station. It follows that the lower the station can be, the more energy may be derived. But we must not be misled by this into erecting the generating plant below sea-level. If we did build a 'submarine' power station we would face the problem of getting rid of the spent water from the turbines. To get rid of it in the sea we would have to pump it upwards, and this would require at least as much extra energy as we have gained.

Fig. 32 (*b*) shows a swinging pendulum at one extreme end of its swing. In this position, it is momentarily at rest. As it swings to the centre, it loses

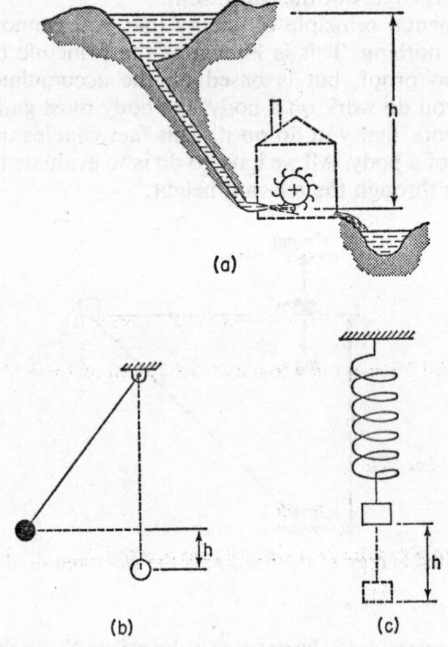

Fig. 32. Three examples of potential energy, showing the height *h*.

potential energy and gains speed. In this case, *h* is the vertical distance between the central and extreme positions of the mass. Fig. 32 (*c*) is similar, except that we now have a mass vibrating on the end of a spring. Here, the height *h* is measured from the lowest point of the vibration, where the mass stops going down and begins to come up.

Now let us go back to the mass in Fig. 31, but instead of applying a vertical force to it, we apply a horizontal force. What will be the result? Assuming no resisting frictional forces, we have a case for the Equation of Motion; the mass will accelerate for as long as we apply the force. We shall suppose that we apply the force for a distance *x* and then remove it, as in Fig. 33.

We have performed work to the amount *Fx*, but we have not increased the potential energy of the body. Instead, we have increased its velocity. A mass having velocity is capable of doing work just as much as a mass raised above a level. We know that the water in the lake in Fig. 32 (*a*) is capable of driving the turbine—but so is the water in a rapidly moving river. Energy possessed by virtue of a body's velocity is called **kinetic energy.** We now require a formula for calculating the kinetic energy of a body in terms of its mass and its velocity.

Referring to Fig. 33, we will suppose that after applying the force F for a distance x, the velocity increases from zero to v. From equation (3) on page 14 we get

$$v^2 = u^2 + 2ax$$

$$v^2 = 0 + 2ax$$

Therefore

$$x = \frac{v^2}{2a}$$

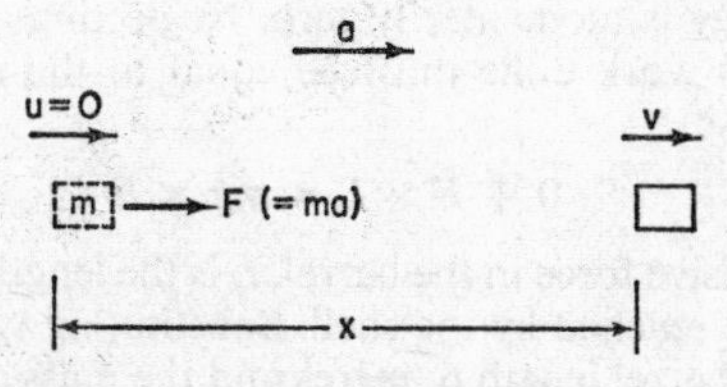

Fig. 33. A horizontal force applied to a mass causes an increase of velocity and hence of kinetic energy.

The Equation of Motion is:

$$F = ma$$

The work done is Fx, which is the kinetic energy J_{kin}

$$J_{kin} = Fx$$

$$= (ma)\left(\frac{v^2}{2a}\right)$$

$$= \tfrac{1}{2}mv^2$$

This formula enables us to calculate kinetic energy directly in terms of mass and velocity. For example, a train of mass 300 tonnes travelling at 100 kilometres per hour will have kinetic energy given by:

$$J_{kin} = \tfrac{1}{2} \times (300 \times 10^3)\left(\frac{100 \times 10^3}{3600}\right)^2$$

$$= 116 \times 10^6 \text{ joules}$$

This is quite a lot of energy. In terms of potential energy, it would be sufficient to raise the whole train vertically approximately 40 metres. Imagine doing this work by hand! And when the train has to be brought to rest, all this energy must be dissipated in the brakes, except for a small amount expended against the air. It comes as no great surprise that train crashes are major disasters.

The principle of conservation of energy may be stated in a kind of equation in words, thus: 'The energy possessed by a body originally, plus any work done *on* the body by some external source of energy, minus any work done *by* the body against any resisting force equals the energy possessed at the end'.

We have tended to simplify many of our examples up to now by neglecting friction, air resistance and similar forces. But these forces usually involve work done *by* a body against some resistance. So, if we continue to neglect them in calculations concerning energy, our answers will be on the optimistic side. To illustrate this point, let us go back to the shell fired vertically from the cannon (page 49). Here is a problem we can readily solve by energy. We do work on the shell in the gun barrel. This work is the product of force and distance. The shell then leaves the barrel with kinetic energy. In ascending vertically, the kinetic energy is converted to potential energy, which will have a maximum value when the shell reaches the top of its flight, and the kinetic energy is momentarily zero. Neglecting any friction, we can say that the original work done must be equal to the maximum potential energy. Algebraically:

$$0 + F \times L = mg \times h$$

where F is the propulsive force in the barrel, L is the length of the barrel and h is the greatest height reached by the shell. Substituting values (the force was 3 meganewtons, the barrel length 6 metres and the mass of the shell 200 kg):

$$0 + (3 \times 10^6) \times 6 = 200 \times 9{\cdot}81 \times h$$

giving a value for h of 9170 metres. So we have obtained the same result as before, but much more easily, and without evaluating the muzzle velocity.

If we know the average resisting force of the air to the ascending shell, we can still use the principle of energy to solve the problem. On leaving the gun, the shell acts against the resistance of the air, and thus does work on the air, which is manifested as heat. In our energy equation, this is work done by the body, and is thus a negative term in the energy equation. This work is the product of average resistance and height of flight. We can guess at an average resistance of one-quarter the weight of the shell. The modified energy equation now reads:

$$0 + F \times L - R \times h = mg \times h$$

where R is the resistance of the air. Again substituting:

$$0 + (3 \times 10^6) \times 6 - (\tfrac{1}{4} \times 200 \times 9{\cdot}81) = 200 \times 9{\cdot}81 \times h$$

this time giving (as we would expect) a lower value for h of 7310 metres.

Should we require the muzzle velocity of the shell, we merely equate the work done in the barrel to the kinetic energy at the muzzle:

$$0 + F \times L = \tfrac{1}{2}mv^2$$

$$0 + (3 \times 10^6) \times 6 = \tfrac{1}{2} \times 200 \times v^2$$

giving again the same value for v that we obtained in Chapter Four, namely 425 metres per second.

At this stage we must pause and reflect that we have solved a problem using two different methods. We obtained the solution on page 50 by using the Equation of Motion. The equation had to be applied twice because two different force systems were involved. The same problem has been solved here using the principle of energy, and with much less trouble. We have to ask ourselves which method should be used in any particular case. There is no simple answer to this, but as a fairly general rule it can be assumed that the energy approach is the best one if you are interested only in the final state of affairs, and not in the motion itself. For example, a train ascends a slope with an initial velocity, under a tractive force and against a resistance, all of which are known. The velocity at the top of the slope will be obtained most quickly by applying the energy equation, but this will not tell you the acceleration or the time taken. To find these, you must use the Equation of Motion.

Perhaps the most impressive demonstration of this rule is provided by the example of a swinging pendulum shown in Fig. 34. Suppose it is at the end of

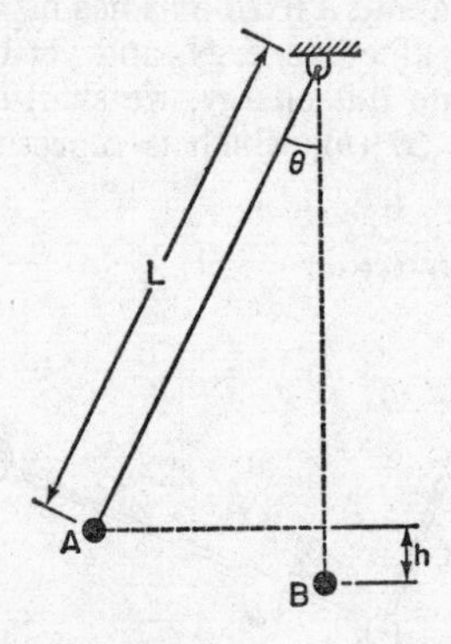

Fig. 34. A swinging pendulum. Potential energy at *A* is converted to kinetic energy at *B*.

its swing in the position *A*, and that we require the velocity at the lowest point *B* of its swing. Assume the length *L* to be 1 metre and the angle θ to be 20°. Since *A* is the outermost point of swing, the mass is momentarily at rest in this position and so has no kinetic energy. Thus the kinetic energy at *B* is entirely obtained from the potential energy lost as the mass descends from *A* to *B*. If the velocity at *B* is *v*, we can say:

$$\text{Loss of P.E.} = \text{Gain of K.E.}$$

$$mgh = \tfrac{1}{2}mv^2$$

$$v = \sqrt{2gh}$$

where *h* is the vertical amount of descent. From simple trigonometry,

$$h = 1 - \cos 20 = 1 - 0{\cdot}94$$

$$= 0{\cdot}06 \text{ metres}$$

$$v = \sqrt{2 \times 9{\cdot}81 \times 0{\cdot}06} = 1{\cdot}085 \text{ metres per second}$$

If we tried to solve this pendulum problem by using the Equation of Motion, we should finish up with a complicated equation called a second-order differential equation, whose solution would require a fair knowledge of advanced mathematics. If you have difficulty in seeing why this should be so for what appears to be a very simple problem, consider the force acting on the mass along the direction of motion. This force is the component of weight along the tangent of the circular path. It varies continuously as the body moves, and so, therefore, will the acceleration. This is just a glimpse of a whole vast field of dynamics called 'vibrations', and having glimpsed it, we shall hastily look the other way! I have merely included it here to show that the use of an energy equation as an alternative to the Equation of Motion may often provide a solution, at the same time avoiding a lot of complicated mathematics.

(3) Kinetic Energy of a Rotating Body

A large wheel spinning about a fixed axis has no absolute linear motion to enable us to evaluate the kinetic energy, and yet there is obviously energy there. In order to calculate this energy, we shall take another look at the system of masses in Fig. 29 (*b*), which is reproduced for convenience in Fig. 35.

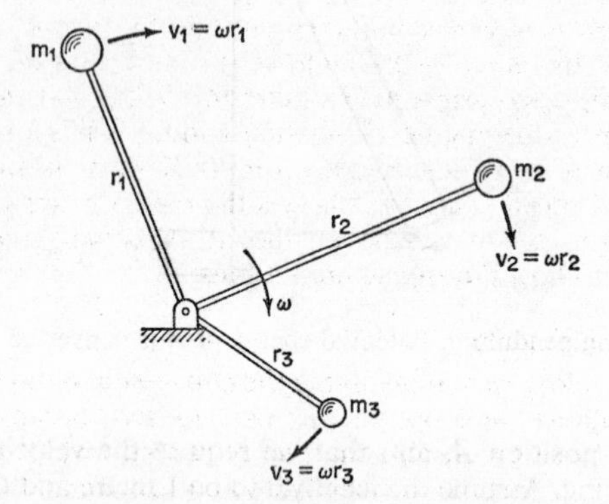

Fig. 35. Each of the separate masses has kinetic energy of $\frac{1}{2}mv^2$.

We assume that the system is spinning about its centre with an angular speed ω. We can calculate the individual linear speeds of the three masses:

$$v_1 = \omega r_1; \quad v_2 = \omega r_2; \quad v_3 = \omega r_3$$

The total kinetic energy is therefore:

$$\begin{aligned} E_{\text{rot}} &= \tfrac{1}{2}m_1v_1^2 + \tfrac{1}{2}m_2v_2^2 + \tfrac{1}{2}m_3v_3^2 \\ &= \tfrac{1}{2}m_1\omega^2r_1^2 + \tfrac{1}{2}m_2\omega^2r_2^2 + \tfrac{1}{2}m_3\omega^2r_3^2 \\ &= \tfrac{1}{2}\omega^2(m_1r_1^2 + m_2r_2^2 + m_3r_3^2) \end{aligned}$$

But we now know from Chapter Four that the quantity in brackets, which can be evaluated for any spinning body, theoretically, or practically, is the moment of inertia I. So our expression for the kinetic energy of rotation is:

$$E_{rot} = \tfrac{1}{2}I\omega^2$$

The flywheel of an engine is essentially a device for storing energy. A petrol engine is driven by pressure forcing a piston down a cylinder. But this action takes place for only a quarter of the time, the remaining time being occupied in scouring out the burnt gases, sucking in fresh fuel, and compressing it. The work done in one stroke has therefore to turn the engine for four strokes. Much of the energy of the power stroke turns the flywheel, which economizes the energy and feeds it back to the engine over the three 'idling' strokes. It would be practically impossible to get a single-cylinder engine, such as a motor-cycle engine, to work without the flywheel. This method of storing energy for subsequent release is also made use of in children's model cars and other 'inertia' toys. The wheels are geared to a small but relatively massive flywheel, so that when the car is started by forcing along the ground by hand, the flywheel releases energy to drive the car for some time after the propelling force has been withdrawn.

Inertia devices are used also for starting the engines of small aircraft. The old-fashioned idea of swinging the propeller of a relatively large and powerful engine is laborious and potentially dangerous. An inertia starter consists simply of a massive flywheel, mounted close to the engine, and arranged for connection to it by means of a clutch, i.e. a connection that can be engaged or disengaged by operating a manual control. When starting, the flywheel is disconnected from the engine. A starting handle is fitted to the wheel, and the operator turns the handle steadily, for 20–30 seconds, until the wheel has attained a good speed. Then he engages the clutch which connects the now spinning wheel to the stationary engine; if all is well, the wheel contains enough energy to turn the engine until it fires.

(4) Energy and Efficiency

In any mechanical system involving the conversion of energy, there always exists something that we loosely term 'loss'; loosely, because the principle of conservation of energy denies that energy can be lost. It can only be converted into another form, or diverted into another channel. But suppose your business is, say, to convert the energy of fuel oil for the purpose of conveying heavy loads by diesel locomotive. In the process of conversion, some of the energy is diverted to heating the engine, which then has to be cooled to prevent it melting or burning, and some more goes to heat the rails the locomotive runs on, and quite a high proportion goes to heat the air, via the exhaust fumes. It is quite reasonable, particularly if you are the one who pays for the fuel, to classify these items under 'loss of energy' because they represent energy not being used directly for the purpose intended. Economy is a vital factor, and it is the business of every practising engineer concerned with this sort of work to keep his losses down to as low a figure as possible, and thus to ensure that the fraction of the original available energy used for the immediate purpose intended is as high as possible. This fraction is called the **efficiency** of the system.

Efficiency is a vital factor in everyone's life. An electrical engineer concerned with providing electrical energy from coal has a public responsibility to make his system as efficient as possible, so that electricity remains relatively cheap to buy. A car manufacturer who can produce a more efficient engine will probably sell more cars because they will travel more miles to the gallon (or more kilometres to the litre). It is a surprising fact, and one that has depressed engineers for decades, that a power station can never achieve an efficiency much in excess of 40 per cent. Some 60 per cent of the energy originally available in the coal is 'lost' or diverted to some other channel where it cannot be employed for producing electricity. In this particular case, most of the 'lost' energy is pumped into the adjacent river in the form of cooling water. There exist no economical means of recovering it. In an atomic energy station, the efficiency is higher; possible efficiencies of as high as 60 per cent are quoted at the time of writing.

The same principle applies when you go to the receiving end, so to speak, and use the electricity for some mechanical purpose. Let us say you propose to operate a lift by means of an electric motor. The energy input can be evaluated from a knowledge of the voltage and current being used by the motor, and the energy output in terms of the mechanical work done. The latter must always be less, although the efficiency in such a case may be as high as 85 per cent. It may be taken as a general rule that, if the employment of energy involves the intermediate stage of heat, the efficiency will be low. The modern engineer is always on the lookout for lost energy and is always taking precautions against it. Modern houses have roof-spaces, and even walls, insulated to prevent loss of useful heat. The temperature of the chimney of a modern house is relatively low, and soot-free. High chimney-temperatures mean loss of heat to the surrounding air, and soot represents lost energy in the form of unburnt carbon. The 'smoke' from a modern power station or factory is practically white, being in fact nearly all steam. The factories of our grandparents' days used to exhibit signs of industry in the form of clouds of black smoke; but black smoke nowadays, even disregarding the health hazard, is recognized as unburnt fuel, and consequent low efficiency.

I have discussed energy in the various forms of chemical energy, heat, electricity and mechanical energy. An important fact to be appreciated is that, in the conversion from one form to another, there is a definite and fixed conversion factor. One kilogramme of coal, or one litre of petrol, will yield just so much heat energy, and no more. The exact quantity is found by carefully-controlled experiment, and is called the **calorific value** of the fuel. Similarly, one unit of heat energy yields its corresponding quota of mechanical energy. This quota used to be called the **mechanical equivalent of heat.** Its value was determined originally by an experiment in which mechanical work was performed on water, causing a rise in temperature, careful precautions being taken against 'loss' of energy. With the rational system of units now being adopted, the same unit of work or energy, the joule, is used for measuring all forms of energy. The joule is our name for the newton metre, and it is also the energy represented by a current of one ampere at a voltage of one volt. The chemical energy of fuel may be quoted as joules per kilogramme or joules per litre directly. With energy in the form of heat, we have to allow a certain degree of flexibility. An alternative unit, the kilocalorie, has to be employed. This is the quantity of heat required to raise 1 kilogramme of water through a temperature rise of 1°Celsius (which is the same as 1°Centigrade). The celsius or centigrade scale is too well established to reject in favour of a rational scale linked to the MKS system, so that the conversion factor is still necessary.

We will close this discussion of energy by a simple calculation of the cost of removing coal from a mine. Let us assume a daily production of 600 tonnes (1 tonne = 1000 kilogrammes), which is hauled up a shaft having a depth of 800 metres by an electric hoist. The hoist is not a perfectly efficient mechanism. Any work done by the driving motor must be more than the work done in raising the load. A reasonable guess at the efficiency of such a hoist would be 90 per cent. The only 'losses' would be friction of the ropes and axles, and a small amount of 'windage' or air friction. The electric motor driving the hoist is also not perfect. Here, we may assume an efficiency (i.e. the ratio of the mechanical work at the output shaft of the motor to the input electrical energy) of about 80 per cent. What will be the daily cost of raising the coal, if electricity costs one penny per kilowatt-hour (100 pence = £1)?

The net daily work performed by the hoist, which we may call J_{out}, is the product of (weight of coal) and (height raised).

$$\begin{aligned} J_{out} &= mgh \\ &= (600 \times 1000 \times 9{\cdot}81) \times 800 \\ &= 47 \times 10^8 \text{ joules or 4700 megajoules} \end{aligned}$$

The electrical energy input will be higher than this figure, and is obtained by *dividing* by the two efficiencies, expressed as fractions, not as percentages.

$$J_{in} = \frac{4700}{0{\cdot}9 \times 0{\cdot}8} = 6530 \text{ megajoules}$$

One kilowatt-hour of electricity is the energy expended by 1000 watts running for one hour. One watt is one joule per second (we shall deal with this in the next section) so that one kilowatt-hour is (1000 × 3600) joules = 3·6 megajoules.

So the daily consumption of electrical energy is:

$$C = \frac{6530}{3{\cdot}6} = 1810 \text{ kilowatt-hours}$$

The daily cost is the same number of pence, or £18·1. If this seems a high figure for a daily electricity bill, remember that the cost *per tonne* is only 3 pence. And imagine how much you would have to pay a gang of labourers to do the same amount of work manually.

(5) Power

If efficiency is one of the most important single factors in modern engineering, then power is certainly of equal rank with it. Power is concerned with the *speed* of doing work. The formal definition of power is 'the rate of doing work'. Its unit is the **watt**, which is one joule per second. Why is power of such overriding importance in modern times? A glance at the example with which we concluded the previous section gives a clue to the answer. A coal mine produces coal at a certain rate. This rate is arrived at by a complicated compromise taking into account the number of employees, the size of the mine, the rate at which coal can be sold, the market price, and a host of other factors. But it becomes clear that at whatever rate the coal is being mined, it must be hoisted to the surface at the same rate if it is not to pile up embarrassingly at the bottom of the shaft. The rate at which such an operation is done is therefore of tremendous importance.

Power is best thought of as a form of currency through which engineers in various fields deal with one another. An engineer designing the coal hoist will specify his requirement of an electric motor in terms of power, not in terms of energy. The choice will lie between a relatively slow, large motor, and a much smaller, high-speed machine. The history of the development of the internal combustion engine during this century shows a trend towards higher speeds, resulting in a general increase of power without a corresponding increase of size. Some of the first engines ever designed were very impressive. Watt's beam engine, still to be seen in the Science Museum in London, gives an entirely false impression of great power. In fact, its actual power was probably less than that of an ordinary saloon-car engine, principally because its speed of operation was so low.

A few simple calculations will give some idea of the order of power in certain situations. A modern car may fairly be expected to ascend a hill of slope one in ten at a steady speed of 60 kilometres per hour. The mass of the car and its load may be guessed at 1·2 tonnes, and the resistance to motion due to road friction and air might have a value of 2 kilonewtons under these conditions. What power is being expended by the car?

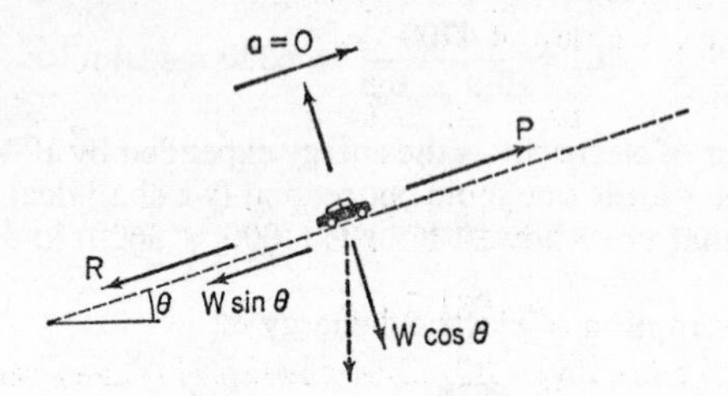

Fig. 36. The free-body diagram of a car ascending a hill.

From the free-body diagram in Fig. 36 you can see that, since the car travels at constant speed, there is no acceleration, and so the total force in the direction of motion must be zero. Thus:

$$P - R - W \sin \theta = 0$$

Sin θ is 0·1 (this is really what a slope of one in ten means); $R = 2000$ N, and $W = mg = 1{\cdot}2 \times 1000 \times 9{\cdot}81$ N. Substituting these values:

$$\begin{aligned} P &= R + W \sin \theta \\ &= 2000 + (1{\cdot}2 \times 1000 \times 9{\cdot}81) \times 0{\cdot}1 \\ &= 2000 + 1180 \\ &= 3180 \text{ newtons} \end{aligned}$$

This force is doing work in moving up the hill. The rate of work, or the work per second, is the product of force and distance moved per second, i.e. the product of force P and velocity v. Calling the power W:

$$W = P \times v$$
$$= 3180 \times \left(\frac{60 \times 1000}{3600}\right)$$
$$= 53\,000 \text{ watts or } 53 \text{ kilowatts}$$

In the old style of power rating this would be about 71 horse-power. Of course the car engine does not drive directly; it acts through a gearbox and transmission system, which would not be perfectly efficient. So for this power at the wheels, a correspondingly greater power would be required of the engine itself—say 70 or 80 kilowatts.

Going back to the coal-hoist of page 75, we calculated that a daily electrical input of 6530 megajoules was required. The power in watts, or joules per second, is thus:

$$W = \frac{6530}{24 \times 3600}$$
$$= 0{\cdot}0755 \text{ megawatts}$$
$$= 75{\cdot}5 \text{ kilowatts}$$

The power capacity of the average human body is rather surprising. We may reasonably suppose that an average healthy man can run up a flight of stairs, ascending a vertical height of say 4 metres in 3 seconds. Taking his mass as 75 kilogrammes, his rate of work is:

$$W = \frac{mg \times h}{t}$$
$$= \frac{75 \times 9{\cdot}81 \times 4}{3}$$
$$= 981 \text{ watts}$$

or, by old-style reckoning, 1·32 horse-power. This may seem flattering, as compared with the humble horse, but the important point to note is that this rate of work is momentary only. For instance, if we demanded that the man run up two flights instead of one, the chances are that the time taken would be at least 8–9 seconds, and the power correspondingly reduced. If we then try and determine the rate at which he could ascend stairs *continuously* for several hours, we should find a drastic reduction in his output, which would probably not exceed 20 watts.

(6) Momentum

This has been called, rather inadequately, the 'quantity of motion' possessed by a body. It is the product of the mass and the velocity of the body. This quantity appears if we examine the Equation of Motion of a body, and substitute for the acceleration the initial and final velocities.

Fig. 37 shows a mass m acted on by a constant horizontal force F for a time t. As a result of this, the mass accelerates from the velocity u to velocity v. The Equation of Motion is:

$$F = m \times a$$

But from the equation $v = u + at$ we can substitute $v - u/t$ for a:

$$F = m\frac{(v - u)}{t} = \frac{mv - mu}{t} \tag{1}$$

The top line of this equation is the change of momentum undergone by the mass. This gives us an alternative way of expressing Newton's Second Law, thus:

force is equal to the rate of change of momentum

For most cases of motion of rigid bodies, this alternative approach does not yield any advantage. But it is often useful in dealing with situations where the change of momentum takes place over an extremely short interval of

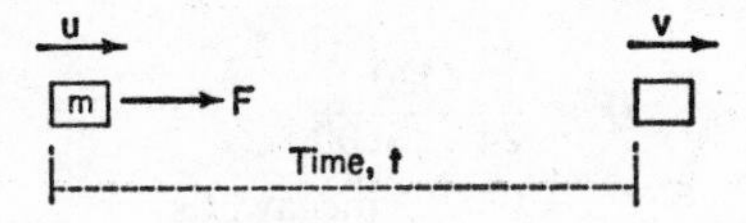

Fig. 37. The mass m, acted on by a force F, increases velocity from u to v in t seconds.

time. In such cases we are unlikely to be able to assess the force itself with any degree of accuracy. Examples are the blow of a hammer, or the impact of a bullet.

In Fig. 38 I have imagined two masses moving in space in such a manner that they collide. To keep the problem as simple as possible, I am assuming that each body moves along the same straight line, both before and after the collision. A simple physical picture would be that of two billiard balls. Before collision, we assume that they are both travelling at steady speeds, u_A and u_B. The result of the collision is to change the speeds to v_A and v_B. During the actual collision there will be some force, which may not be constant, and which I have called f, acting for some indeterminate time t.

The approach to the solution of this problem lies in Newton's Third Law, which I hope you will recall, as action giving rise to an equal and opposite reaction. In this case, at any instant of time during the collision, the force exerted by the mass A upon B must be exactly the same as, but in the opposite direction to, the force exerted by mass B upon A. Furthermore, the time of collision contact is obviously the same for both masses. So we are able to write an equation of motion in the momentum form for each of the two masses, in terms of the unknown force and the time. But we must remember to give one of the forces a negative sign because it is in the opposite direction. We shall assume that the direction from left to right is positive.
Equation (1) may be rewritten:

$$F \times t = mv - mu$$

Substituting values for masses A and B in turn:

$$-f \times t = m_A v_A - m_A u_A$$
$$+f \times t = m_B v_B - m_B u_B$$

Adding the two equations together to get rid of the unknown f and t:

$$m_A v_A - m_A u_A + m_B v_B - m_B u_B = 0$$

If we rearrange this with the u terms on one side and the v terms on the other:

$$m_A u_A + m_B u_B = m_A v_A + m_B v_B \qquad (2)$$

we can see that the left-hand side of the equation is the total momentum of the system before the collision, and that the right-hand side is the total momentum after the collision.

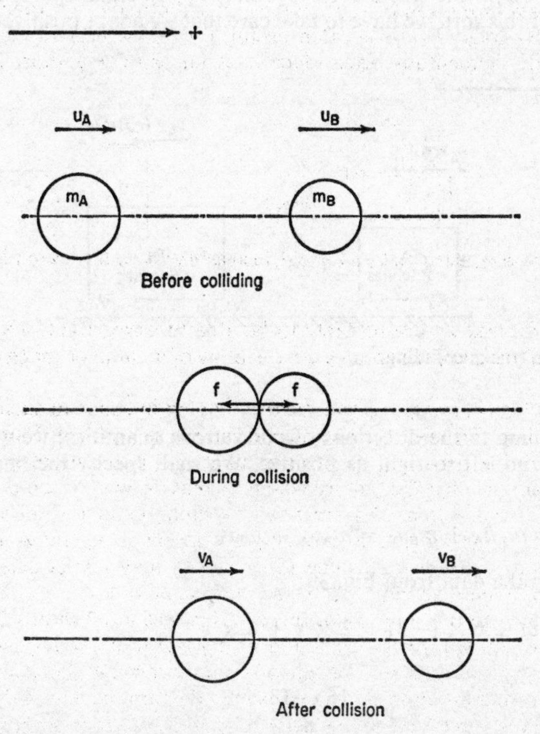

Fig. 38. Two masses colliding. The impact of *A* on *B* is exactly equal and opposite to that of *B* on *A*.

Equation (2) is sometimes described as the 'law of conservation of momentum', because it can be stated in words that the total momentum before impact is equal to the total momentum after impact. Although this is true, it has some unfortunate consequences. First, it is often confused in the minds of students with the *principle* of conservation of energy. I shall shortly show that this is one particular situation of dynamics where the principle of conservation of energy cannot be used. Secondly, although I have not space to illustrate here, there are cases of impact between systems (for example, rotating systems) where momentum is *not* conserved during collision.

I shall conclude by a simple illustration of the use of this equation. Let us consider that item beloved of writers of textbooks on dynamics, two trucks on a rail.

Fig. 39 shows the trucks before collision. The left-hand truck travels to the right at 8 metres per second, and the right-hand truck travels in the opposite direction at 3 metres per second. We assume that on colliding they automatically lock together. We have to make some assumption of this sort, because the momentum equation we have derived contains two unknown velocities v_A and v_B, and one equation cannot yield two unknown quantities. So in this case, if the trucks lock together, v_A will equal v_B. As always with problems of this sort, we have to take care that we adopt positive or negative

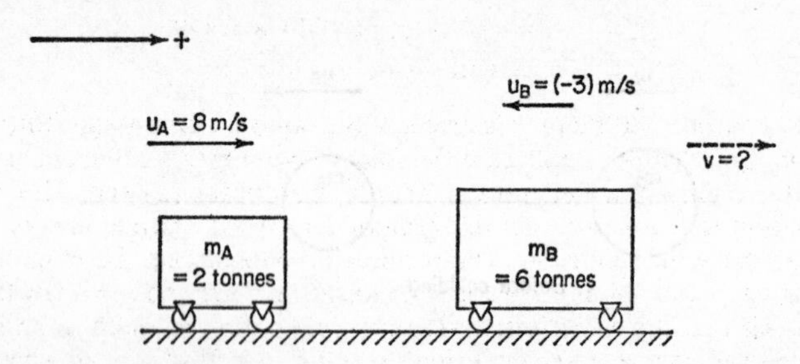

Fig. 39. Two trucks colliding.

signs according to the directions of our various quantities; we will take the direction from left to right as positive. We shall specify the final common velocity as v.

$$m_A u_A + m_B u_B = m_A v_A + m_B v_B$$

Substituting the data from Fig. 39:

$$2 \times (+8) + 6 \times (-3) = 2 \times v + 6 \times v$$
$$= 8 \times v$$

Therefore
$$v = \frac{16 - 18}{8}$$
$$= -0{\cdot}25 \text{ metres per second}$$

Since the answer we obtain is negative, the final common velocity of the two connected trucks must be to the left.

It may be observed that on this occasion I have failed to express all the terms in correct and consistent units. I have left the masses in tonnes instead of converting to kilogrammes. But you can see that such a conversion would have cancelled out in the equation. Similarly, if the speeds had been given in kilometres per hour instead of metres per second, we could have left them in that form, and the answer obtained would have been in the same units.

The most important lesson to learn from this example concerns the energy. A common error in the solution of this type of problem is to equate the total

kinetic energy before collision to the total kinetic energy after collision. Let us evaluate the two quantities now.
The total kinetic energy before collision is:

$$\begin{aligned} E_1 &= \tfrac{1}{2}m_A{u_A}^2 + \tfrac{1}{2}m_B{u_B}^2 \\ &= \tfrac{1}{2}(2 \times 1000)(8)^2 + \tfrac{1}{2}(6 \times 1000)(-3)^2 \\ &= 64\,000 + 27\,000 \\ &= 91\,000 \text{ joules} \end{aligned}$$

The total kinetic energy after collision is:

$$\begin{aligned} E_2 &= \tfrac{1}{2}(m_A + m_B)v^2 \\ &= \tfrac{1}{2}(2 + 6) \times 1000 \times (0{\cdot}4)^2 \\ &= 640 \text{ joules} \end{aligned}$$

So that you can see there is a tremendous amount of energy 'lost' in this example. To attempt to calculate the final velocity by equating energy before and after collision would give a grossly inaccurate answer. The physical explanation is that a considerable proportion of the original energy is dissipated at the actual collision. The proportion depends on the magnitudes of the original velocities and masses. For instance, if we took two trucks of the same mass and the same initial velocities, and allowed them to collide and lock together, they would each bring the other to a dead stop; in such a case, *all* of the original kinetic energy would be lost to the system.

CHAPTER SIX

FRICTION

(1) Dry and Sliding Friction

Friction is a force encountered in all moving machinery, and comprises a resistance to sliding motion. Essentially, the cause is the relative roughness of the sliding surfaces. Friction may be either an enemy or a friend. Friction in rapidly moving machinery, if unchecked, would give rise to heat which would seriously damage the parts in a very short time. A considerable amount of design ingenuity is devoted to minimizing or eliminating these unwanted forces. In a modern car, the complicated lubrication system has the dual function of keeping frictional forces to within acceptable limits, and of dispersing the heat generated by these forces. If it were not for this lubrication, the heat generated at some of the sliding surfaces (perhaps the pistons in the cylinders) would be so great as to cause local heating, and consequent fusion of the surface metal, and the engine would 'seize up'. Any good driver keeps a frequent watch on his fuel gauge, but his oil-pressure gauge deserves even closer and more frequent attention. An empty tank can soon be filled, but a damaged bearing or cylinder calls for a major overhaul.

On the other hand, there are at least four instances in the modern car where friction is exploited; indeed, in one of these, it is difficult to visualize how we could do without it. The propulsion of the car along the road relies entirely on friction between tyres and road surface. The choice of material for the tyre, and the design of the pattern, or tread, is determined largely by a consideration of the maximum friction it is possible to create. The clutch, connecting the engine to the wheels, and the belt driving the dynamo, are also examples of the use of friction, although in these cases alternative methods could be used. Finally, the brakes reduce the speed of the car by applying a friction force to the inside of a drum attached to the wheel. Quite a high proportion of the kinetic energy of the moving car is used in doing work against friction at the brake drums, with the result that there is a considerable rise of temperature of the drums after heavy braking.

Before going on to consider the more theoretical aspects of this topic, there are two important points we have to appreciate. The first concerns the use of friction for beneficial purposes, as in the examples above. In such cases, the logical procedure would appear to be to obtain the maximum possible force of friction. For instance, in designing a belt drive in which a flat or V-shaped leather or composition belt connects two pulleys, the pulley surfaces could be coated with emery to 'bite' the belt. But in practice this is never done. The rubbing surface of such a pulley will always be found to be almost glass-smooth. The reason is that a compromise has to be made between frictional force and wear. With a roughened pulley, the belt would be worn away in a very short time. Similarly, rough roads will always give a better grip than smooth ones, but they will also wear out the tyres quicker.

The second point to note concerns the lubrication of friction surfaces. We shall find in the following section that the laws of friction are not so reliable as, for instance, Newton's Laws of Motion, or Hooke's Law. One reason is that frictional forces arise out of conditions which it is either very difficult or impossible to define accurately.

If we take two blocks of steel which have been planed on a machine, and examine the friction between them, we shall find that this force depends on several factors, some of which we shall be unable to specify. A planing machine cuts horizontal strips at closely spaced intervals along a flat piece of metal, rather in the manner of ploughing a field. Although the planed surface is nominally flat, it actually consists of ridges and furrows due to the action of the cutting tool. The friction between two such similar surfaces will depend on (*a*) the shape and sharpness of the cutting tool, and (*b*) the rate of sideways feed or 'traverse' of the tool, i.e. the 'pitch' or spacing of the adjacent ridges and troughs. Also, the friction must depend on *how* the surfaces are offered together. If they are placed with the grooves and ridges in line, the frictional force will be relatively small if the sliding is parallel to the grooves. But if the motion of sliding is across the line of the grooves, the frictional force will be large. In the first case, ridges and grooves will tend to fit together and slide one within the other; in the second, there will be the maximum resistance to motion as ridges on one surface have to 'jump over' ridges on the other.

Finally, the friction might get progressively less as the surfaces rub together for a time, and the more prominent ridges become worn down. This is made use of in the engineering process known as 'lapping', in which mating sliding surfaces are made very smooth by rubbing together with a fine grinding compound between. The same process takes place in the 'running-in' of an engine. A by-product of such a process is the 'swarf', which consists of the tiny particles of metal rubbed from the surfaces. These are carried off by the lubricating oil. Thus an essential sequel to running-in a machine is the changing of the oil, so that the particles of swarf cannot be carried round to damage the newly smoothed surfaces.

In spite of all these rather random factors, it is found that certain generalizations can be made in practice, resulting in the formulation of what are called the Laws of Dry Friction. It will be our task to examine these laws in the section which follows. But the introduction of lubricating oil between two sliding surfaces completely alters the conditions. This is because the function of lubrication is actually to *separate* the two surfaces by a thin film of oil. Consequently, when sliding occurs, the force of friction is due only to the shearing of this film of oil, and not due to actual contact between surfaces. The best illustration of this is the action of a car wheel on a snow-covered road. The spinning of the wheel actually drives a mixture of snow and water between the wheel and the road, so the tyre is prevented from making contact with the road by a thin film of melted snow, less than one-thousandth of an inch thick. This, of course, is a disadvantage; but the same action is exploited in the lubrication of shafts turning in bearings, and in all parts of machines where relative sliding would otherwise give rise to excessive heat.

The frictional forces associated with lubricated surfaces are thus quite different from those on dry surfaces, and the laws governing them are different. Furthermore, the study of these forces is far more complex, and a whole field of study known as **tribology** is now devoted to them. In what follows, we shall only be concerned with the very simple laws associated with sliding of dry, unlubricated surfaces. These laws have been known for many ages, and although they can only be considered as approximate, an investigation will serve to show some interesting aspects of the subject.

(2) The Laws of Dry Friction

The first law of dry friction states that the frictional force between sliding surfaces is *proportional to the normal force between the surfaces.* With reservations, this may be considered the most useful, and the most accurate law. In practice, it tells us that the horizontal force required to move, say, a loaded table along a horizontal floor, will be doubled if the load on the table (including its own weight) is doubled. Simple experiment verifies this law, and many readers may recall performing such an experiment by placing weights on a wooden block which slides along a fixed surface.

The second law states that the frictional force is *independent of the area of contact between the sliding surfaces.* This is true up to a point, but is far more dubious and open to challenge. If the law is true, we may turn the loaded table of our example upside down, whereupon the same force should be necessary to move the same load. This may well be found to be so. But if we fit spikes to each leg of the table, it becomes clear that the law would not be valid for such a case. This law, then, must be applied with discretion.

The third law states that frictional force is *independent of the velocity of sliding*. Again, there are reservations. It is found in practice that the force required to *start* two surfaces sliding relative to each other is always greater than the force required to *keep* them sliding. The former force is called **static friction** (rather engagingly abbreviated to 'stiction') and the latter is called **dynamic friction.** So if we limit this law to dynamic friction, we may accept it as valid.

The fourth law cannot be considered as a law at all; it merely states the obvious fact that frictional force depends on the nature of the surfaces, without giving us any further clue as to how to determine the force.

At no point do these four laws help us to calculate frictional forces. If our business is to predict the frictional force between a loaded steel shaft and a brass bearing surface, we shall either have to perform a test on a pair of mating surfaces as nearly identical as possible to those we are interested in, or we shall have to rely on the results of a similar test undertaken in the past by someone else. Even then, bearing in mind all the possible variable factors, some of which I have outlined, we must not expect our prediction to be much closer than within 20 per cent of the true value.

(3) Coefficient of Friction

With these reservations in mind, let us now, in imagination, perform a simple experiment on the apparatus shown in Fig. 40 (*a*). A wooden block is caused to slide along a wooden surface by means of a cord carrying weights. The load on the block may be varied.

If the first law is true, we should find that the frictional force, as determined by the weight w in the pan, increases in proportion to the normal force between the surfaces, as determined by the weight W on the block (assuming that W includes the weight of the block itself). The relationship would be *linear* (see page 43), and a graph of w against W should have the straight-line form shown in Fig. 40 (*b*). In practice, this would be found to be so. If the wooden block were exchanged for a steel block, or the wooden table replaced by a glass surface, we should still obtain a straight-line graph, although not the same one. If on the other hand we turned the wooden block over, increasing the area of surface contact, the second law tells us that this should not affect the result, and within the rather broad limits of possible experimental error, this also would be found to be so.

Such a graph as we obtained would represent a constant factor for a pair of sliding surfaces (in this case, wood on wood). Using the same materials,

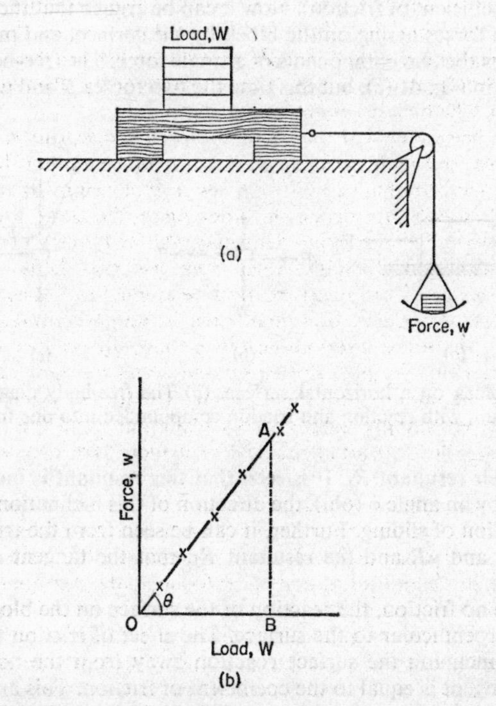

Fig. 40. (*a*) Apparatus for measuring friction force for varying loads. (*b*) Graph of frictional force against normal force.

we should always get the same straight line, and the relation between the normal force and frictional force could be expressed by the slope of this straight-line graph. This would be the tangent of the angle θ in Fig. 40 (*b*), or the fraction AB/OB. This fraction is called the **coefficient of sliding friction** for the particular pair of sliding surfaces, and is usually designated by the Greek letter μ (mu).

For any frictional force to be calculated, then, the value of μ for the appropriate sliding surfaces must be known—either as the result of direct experiment, or by relying on the result of experiment by someone else. Approximate values of μ are quoted for many mating surfaces in books on engineering and mechanics, and these are useful to an extent depending on the conditions.

For instance, 'wood on wood' is so vague as to be virtually useless, whereas 'steel on brass' may well serve to calculate the force necessary to operate a screw-jack with moderate accuracy.

(4) Angle of Friction

Examine the single block resting on the horizontal surface, in Fig. 41 (*a*). The free-body diagram of the block is shown in Fig. 41 (*b*), and it is seen that four forces act upon the block: the force F, the weight W, the induced reaction R of the surface, and the induced frictional force (which must be the product of R and the coefficient of friction). Now it can be argued that these two latter forces are both forces acting on the block by the surface, and may therefore be considered as the two components of a single force. The free-body diagram is shown again in Fig. 41 (*c*), but this time the two forces R and μR have been

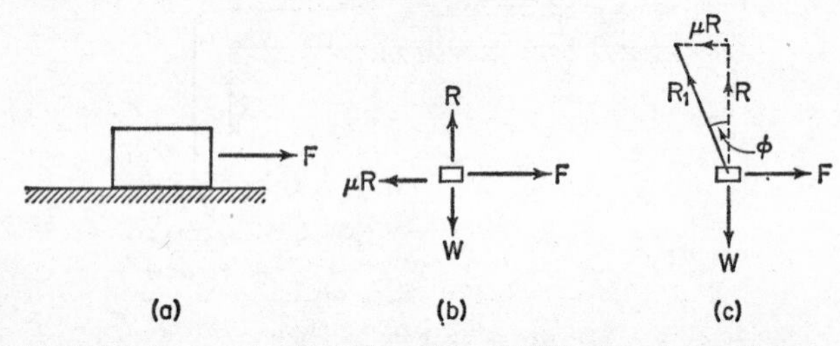

Fig. 41. (*a*) A block on a horizontal surface. (*b*) The free-body diagram. (*c*) The simplified diagram, with reaction and friction compounded into one force.

replaced by their resultant R_1. It is seen that this resultant is inclined to the perpendicular by an angle φ (phi), the direction of this inclination being away from the direction of sliding. Further, it can be seen from the triangle of the components R and μR and the resultant R_1, that the tangent of the angle φ is $\mu R/R$ or μ.

If there were no friction, the reaction of the surface on the block would be normal, i.e. perpendicular to the surface. The effect of friction may thus be considered as inclining the surface reaction away from the normal by an angle whose tangent is equal to the coefficient of friction. This angle is called the **angle of friction**. We shall see in some of what follows that the use of the angle of friction often simplifies the solution of friction problems. As an illustration of the type of simplification I am talking about, consider the situation illustrated in Fig. 42. Here we have once again the simple case of a block of weight W resting on a flat horizontal surface, but instead of moving it by a horizontal force, we pull it by a force P which is inclined upwards at an angle θ. At what value of θ will the force P have its least value?

In explanation, it must be emphasized that the frictional force varies with the normal force between surfaces, not necessarily with the weight of the body. The effect of inclining the force P upwards will be to reduce this normal force, and hence to reduce the force of friction.

Fig. 42 (*b*) shows the free-body diagram, with the surface reaction now represented by the single force R_1 inclined to the normal at angle φ, the angle of friction. In order to examine how the variation of angle θ will affect the

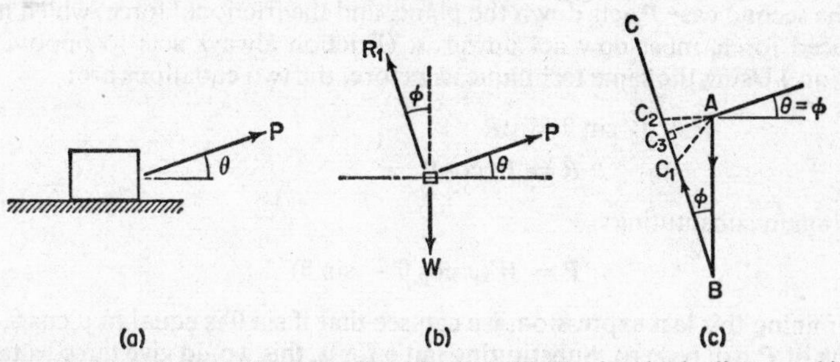

Fig. 42. (*a*) A block pulled by an inclined force *P*. (*b*) The free-body diagram. (*c*) The triangle of forces.

value of *P*, we draw the triangle of forces as shown in Fig. 42 (*c*). We start by drawing force *W*, because this must always be downwards and of constant magnitude. This is shown as *AB*. Secondly, we can draw a line of indeterminate length, at an angle φ to the vertical, to represent force R_1. This is the line *BC*. The closing line of the triangle must pass back through *A*. Various possible positions of this line are shown dotted as AC_1, AC_2, AC_3. It is clear from this diagram that the least possible value of force *P* (which is proportional to the length of the side of the triangle) is that given by the line AC_3 where the angle at C_3 is a right-angle. The corresponding angle θ for this case is clearly φ.

(5) The Inclined Plane

A fair amount of insight into friction may be obtained by considering various problems in which a block rests on a surface inclined at an angle θ to the horizontal. Let us specify *W* as the weight of the block, and *P* as the force to move it in the required direction. The coefficient of friction may be represented by the angle of friction φ in each case. With this in mind, let us now solve four simple problems. With the block resting on the plane, let us obtain expressions for the value of the force *P* acting parallel to the plane, first, to pull the block up and, secondly, to pull it down. Then let us repeat the problems, with the force *P* horizontal instead of parallel to the plane. We shall gain more insight if we solve these problems in algebraic terms rather than making arithmetical substitutions.

Fig. 43 shows the arrangement when the force *P* is parallel to the plane. In this particular case, there is no advantage in using the angle of friction, and so the four forces *P*, *W*, *R* and μR are shown separately. Solution is obtained by writing equations of equilibrium in two directions. These are most conveniently chosen as the directions parallel to, and perpendicular to the plane. The same result could be obtained by writing equations along the horizontal and the vertical directions, but the algebra would be slightly more complicated. The two equations resulting from the first case are:

$$P = \mu R + W \sin \theta$$
$$R = W \cos \theta$$

Substituting the value of *R* in the first equation:

$$P = W(\mu \cos \theta + \sin \theta)$$

In the second case P acts down the plane, and the frictional force, which is an induced force, must now act upwards. (Friction always acts to oppose the motion.) Using the same technique as before, the two equations are:

$$P + W \sin \theta = \mu R$$
$$R = W \cos \theta$$

and again substituting:

$$P = W(\mu \cos \theta - \sin \theta)$$

Examining this last expression, we can see that if $\sin \theta$ is equal to $\mu \cos \theta$, the value of P will be zero. Substituting $\tan \varphi$ for μ, this would give $\tan \varphi = \tan \theta$,

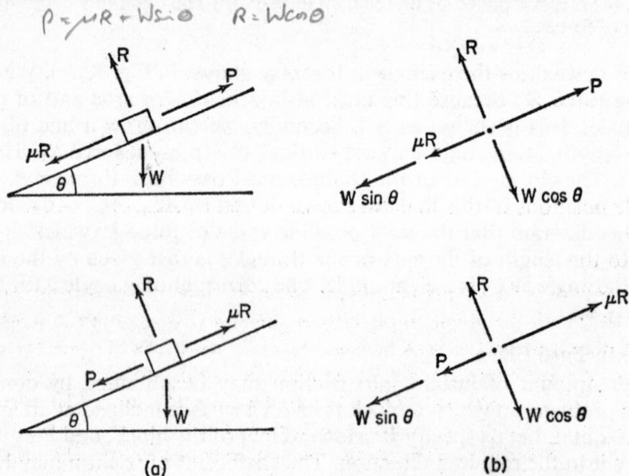

Fig. 43. The inclined force to move a block on a plane. At (*b*) is shown the force system with W resolved into components parallel to and perpendicular to the plane.

or $\varphi = \theta$. In other words, the block would need no force to pull it down the plane if the angle of inclination of the plane were the angle of friction. This gives us a simple physical meaning of the angle of friction: it is the angle of inclination of a plane such that a block would just slide freely down it.

The last two of our four problems are illustrated in Fig. 44. This time, it is more convenient to use the angle of friction to obtain a quick solution. In the first case, since motion takes place up the plane, the reaction R_1 is inclined to the normal to the plane downwards. The triangle of forces is shown alongside, and it is now seen to be a simple right-angled triangle, having the upper angle of magnitude $(\varphi + \theta)$. Hence:

$$P = W \tan (\varphi + \theta)$$

A similar treatment of the fourth problem gives the similar result:

$$P = W \tan (\varphi - \theta)$$

and again it is seen that if $\theta = \varphi$ the value of P will be zero.

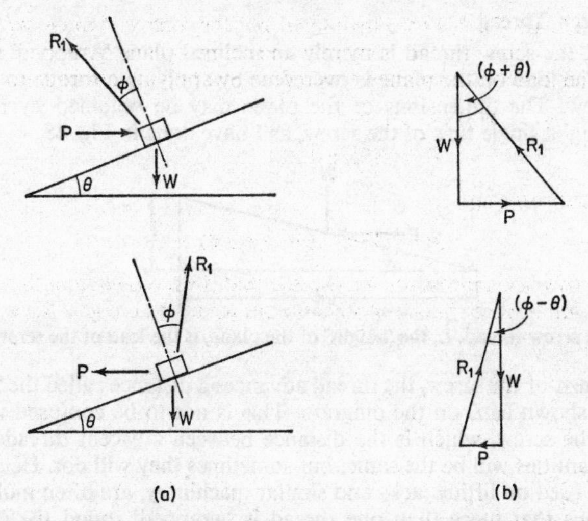

Fig. 44. The horizontal force to move a block. At (*b*) is shown the triangle of forces.

(6) Efficiency

The principle of the inclined plane is made use of in several practical devices, of which the most important is the screw thread. In a screw jack, work is performed on the jack by the operator, and work is done by the jack in raising the load. Since friction is present, there must be some 'loss' of work or energy and the jack will therefore not be fully efficient. The theoretical efficiency can be calculated in terms of the coefficient of friction or the angle of friction. We have already obtained the expression for the horizontal force P to raise the load W up the inclined plane:

$$P = W \tan(\varphi + \theta)$$

If the force of friction were zero, the plane would be 100 per cent efficient, and the angle of friction φ would be zero. Then the theoretical force P_t to raise the load would be:

$$P_t = W \tan \theta$$

so that a simple expression for the efficiency of the plane is the ratio of the theoretically perfect value P_t to the actual value P.

$$\text{Efficiency} = \frac{P_t}{P}$$

$$= \frac{W \tan \theta}{W \tan(\varphi + \theta)}$$

$$= \frac{\tan \theta}{\tan(\varphi + \theta)}$$

You can see from this final expression that if φ is zero (i.e. when there is no friction) the fraction $P_t/P = 1$ and the efficiency is 100 per cent.

(7) The Screw Thread

In effect, the screw thread is merely an inclined plane 'wrapped' round a cylinder. The load on the plane is overcome by applying a torque to the axis of the screw. The dimensions of the plane may be obtained by mentally 'unwrapping' a single turn of the screw, as I have done in Fig. 45.

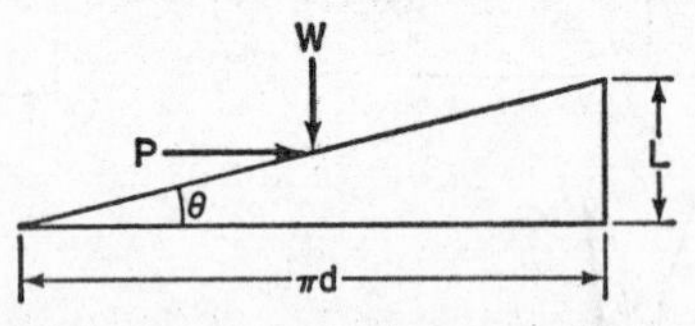

Fig. 45. The screw thread. *L*, the 'height' of the plane, is the lead of the screw.

In one turn of the screw, the thread advances a distance called the 'lead' of the screw, shown as *L* on the diagram. This is not to be confused with the 'pitch' of the screw, which is the distance between adjacent threads. Often the two quantities will be the same, but sometimes they will not. Heavy-duty threads, as used on lifting jacks and similar machinery, are often multi-start, which means that more than one thread is 'wrapped' round the cylinder. Fig. 46 illustrates a two-start thread, and it is seen that the lead, or advance per turn, is twice the pitch.

The required torque to operate the screw is determined by taking the moment of the driving force *P* about the centre of the screw. The effective moment arm of this force is half the mean screw diameter.

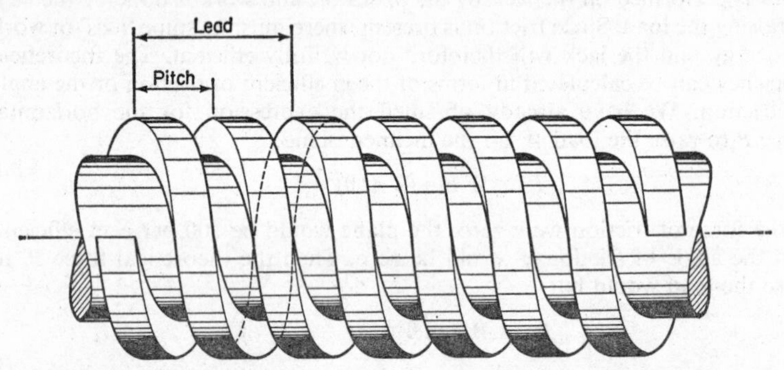

Fig. 46. A two-start thread.

As a simple exercise, consider a simple screw jack such as might be used for raising a car. Assume a mean screw diameter of 2 centimetres, and a lead of 1 centimetre. These dimensions would give reasonable proportions for a screw; the angle of thread of a screw jack is always much coarser than that of a screwed bolt. Taking a possible coefficient of friction of 0·1, let us determine the required torque on the jack to raise a load of 10 kilonewtons.

Referring back to Fig. 45, the angle of the plane is given by

$$\tan\theta = \frac{L}{\pi d} = \frac{1}{\pi \times 2} = 0{\cdot}159$$

giving a value of 9° for θ. φ is the angle whose tangent is μ, which is 0·1 in this example, giving $\varphi = 5{\cdot}7°$. The case corresponds exactly to the third of the four problems we considered in Section 5 and we may use the equation on page 88 to determine the required operating force P.

$$\begin{aligned} P &= W \tan(\varphi + \theta) \\ &= 10 \tan(9 + 5{\cdot}7) \\ &= 10 \tan 14{\cdot}7 \\ &= 10 \times 0{\cdot}262 \\ &= 2{\cdot}62 \text{ kilonewtons} \end{aligned}$$

The torque required on the screw is the product of this force and the mean screw radius:

$$\begin{aligned} \text{Torque} &= 2{\cdot}62 \times (1 \times 10^{-2}) \\ &= 0{\cdot}0262 \text{ kilonewton metres} \\ &= 26{\cdot}2 \text{ newton metres} \end{aligned}$$

Such a torque might be provided by a force of 131 newtons (approximately 30 pounds) at the end of a lever arm of radius 20 centimetres. In a more complicated design of jack, the torque might be provided through the medium of a rotating handle and a bevel gear.

Assuming that the jack is a simple screw thread only, we can use the expression on page 89 to determine the theoretical efficiency:

$$\begin{aligned} \text{efficiency} &= \frac{\tan \theta}{\tan(\varphi + \theta)} \\ &= \frac{0{\cdot}159}{0{\cdot}262} \\ &= 61 \text{ per cent} \end{aligned}$$

This is far too high a value for the efficiency of an actual jack, for a reason that I shall discuss in the following section. It should be fairly clear from the theoretical expression that the efficiency will increase as θ becomes greater, and this is generally so. But with an apparatus such as a lifting jack there is another very important consideration to bear in mind: the jack must not be reversible, in the sense that the load can cause it to run down. As we shall see, one consequence of this requirement is that the efficiency must not be too *high*. For simple manually operated jacks, efficiency is not of great importance in comparison with the main task of raising a certain load without allowing it to 'overhaul', i.e. letting the jack descend under the load without manual operation. In more elaborate jacks, where efficiency might be important, friction is minimized by the use of a precisely manufactured steel screw running in a well lubricated bronze nut, while the problem of overhauling is solved by driving the screw mechanically and providing automatic locks or brakes. A typical example is the operating screw on a power-operated tipping lorry.

(8) Reversibility

The problem of overhauling, which we have briefly touched upon, can be investigated by examining the conditions under which a theoretical lifting machine may be driven backwards by the weight of the load. Such a machine may be described as **reversible**. The problem is really outside the range of simple screw threads, because in most lifting devices there are other sources of frictional energy dissipation. For instance, even in the simplest jack, the collar or nut carrying the load usually rests on a fixed frame, and quite a high proportion of the torque is required to overcome the friction at this point, quite apart from the friction of the screw itself. This is one reason why an efficiency of 61 per cent for a jack was described earlier as unrealistic.

Let us now imagine a lifting machine reduced to its simplest elements: a box, with a platform for raising a load W, and a handle for supplying an operating force P. We may assume that the operating force P moves a distance x in raising the load W a height y. The ratio x/y is called the **velocity ratio** of the machine. Now the force to raise the load may be called P_R, and that to lower the load is P_L. What we have to find is the necessary condition for P_L to be zero. At this stage, the load would be just sufficient to cause the jack to overhaul. Let us first raise the load and then lower it; we will write an energy equation for each stage. We must assume in each case that a certain amount of energy is 'lost', or dissipated in friction. We shall call this quantity of energy F, and shall assume that it is the same amount in raising the load as in lowering. (This will not be strictly true, but is accurate enough for our purpose.)

In raising the load, we start off with zero energy. Work is done *on* the machine by P_R, and is done *by* the machine against friction. The final energy is the potential energy of the raised load W. The energy equation is therefore:

$$0 + (P_R \times x) - F = W \times y$$

In lowering the load, we start off with the same potential energy. We do work *on* the machine by P_L. Work is done *by* the machine against friction. The final energy is zero. The equation is:

$$(W \times y) + (P_L \times x) - F = 0$$

If we eliminate F from this pair of equations by subtracting the second from the first, we get:

$$P_Rx - Wy - P_Lx = Wy$$
$$(P_R - P_L)x = 2Wy$$

The efficiency of the machine when raising the load is given by the fraction of the work done *by* the machine over the work done *on* it:

$$\text{Efficiency} = \frac{Wy}{P_Rx}$$

Now if the machine is reversible, the value of P_L must be zero. So our equation $(P_R - P_L)x = 2Wy$ has to be modified to:

$$P_Rx = 2Wy$$

from which it can be seen that $Wy/P_Rx = 0{\cdot}5$.
This tells us that if the efficiency of a machine is more than 50 per cent, the machine will be reversible—and the load will be capable of driving the machine backwards.

(9) Some Examples of Friction

We shall conclude this discussion on friction by examining one or two applications of the principles we have established. First, let us examine the case of a person climbing a ladder which rests against a wall. For our purpose, we shall assume a flat ground and wall, and that the friction coefficient is the same for both. The situation is shown in Fig. 47.

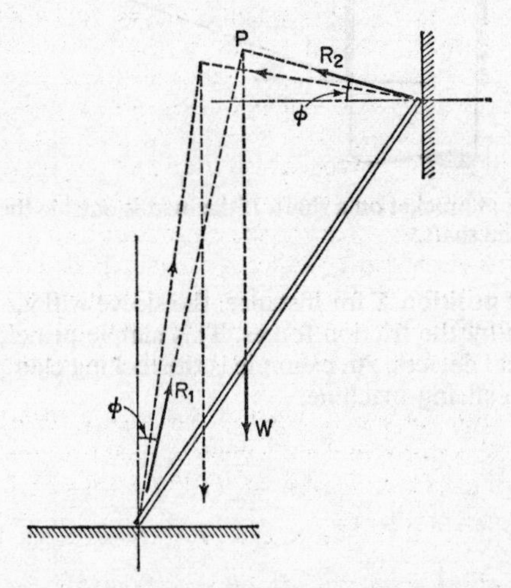

Fig. 47. A man mounting a ladder. For equilibrium, all three forces must pass through one point.

If we assume that the ladder is about to slip, the reaction forces of the ground and the wall upon the ladder will be inclined as shown, at an angle φ away from the normal and in the direction of slip. These two reaction forces R_1 and R_2 will meet at the point P.

For equilibrium, the line of action of the man's weight W must also pass through P. If he attempts to climb higher up the ladder, it will slip. If he is lower down the ladder, the frictional forces will be less than the maximum values, and the line of action of the three forces will probably be roughly as shown dotted. The practical lesson to learn here is that the fact that the ladder does not slip when you are at the bottom, is no guarantee that it will not do so when you are at the top.

Consider now the situation shown in Fig. 48. Here we have a sleeve which is a fairly close sliding fit on a vertical shaft, and a bracket is attached to the sleeve. The bracket supports a load, which may, or may not cause the sleeve to slide along the shaft. The load will cause a slight turning of the sleeve, such that it will contact the shaft at the top-left and bottom-right corners only. Again, the reactions of the shaft on the sleeve at these points will be inclined to the normal at the angle of friction, in the direction of relative sliding. These two reactions will meet at the point P. If the line of action of the load W is to the left of P, sliding of the sleeve along the shaft will occur; on the other hand,

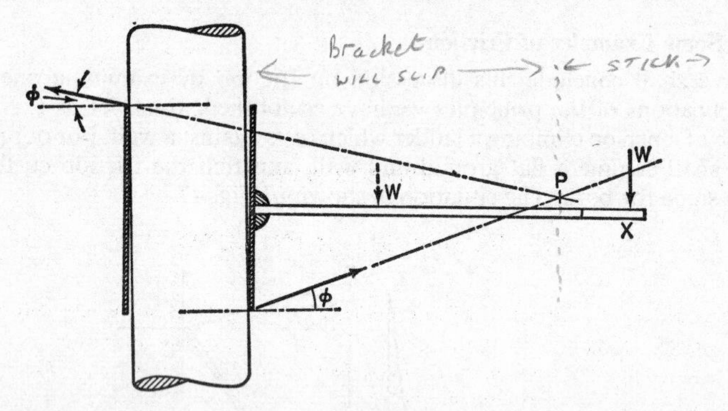

Fig. 48. A sleeve and bracket on a shaft. If the load is outside the point P, the sleeve will not slide on the shaft.

if it is outside, at position X for instance, the sleeve will jam on the shaft and will be held firm by the friction forces. This simple principle is exploited in several mechanical devices. An example is the locking clamp used to hold the joint on a bacon-slicing machine.

CHAPTER SEVEN

FORCES IN SIMPLE FRAMEWORKS

A considerable amount of engineering construction work consists of what are called 'steel-frame structures'. These are frameworks assembled from standard steel bars of various cross-sections. Examples vary from a very simple triangular framework of only three bars, for the support of a small roof, to the complex mass of members and girders comprising a large steel bridge. An obvious requirement for the design of such a framework is that the load sustained by each component member shall be known, and in this chapter I shall describe certain methods of determining these loads.

To evaluate the loads within a structure by using the methods of statics implies the assumption that the structure is at rest while loaded, or that such motion as it may have will not affect the loads in the members. Broadly speaking, this is a reasonable assumption to make. If we consider an obviously stationary structure, such as a bridge, it may be stated with confidence that the use of statical methods to calculate the loads in the members will give an accurate result. Although some of the load carried by the bridge is moving, and we may have to calculate the effects of this movement using dynamical methods, we can take the result of the calculation and apply it to the bridge as a static load. For example, the movement of a truck across the bridge may occasion a maximum load of three or four times the weight of the truck. But having determined this magnified load, we may then design the bridge to take this load, using the methods of statics.

For a structure which moves, such as a crane, the normal movement will not give rise to high loads within the structure *unless this movement consists of high accelerations of the structure itself*. Such a contingency could arise, for instance, if an operator suddenly reversed the horizontal swing of the boom of a large tower crane. Since this boom has a mass of many tonnes, its sudden reversal would require inertia forces which might well be sufficient to damage the structure. The contingency is extremely remote, because the designer of the crane would almost certainly include an overload device, so that any attempt on the part of the operator to perform such light-hearted gyrations would automatically switch off the driving motor. Nevertheless, the existence of possible dynamic loads in a static structure must always be borne in mind.

History can quote at least two cases of failure of structures due to the presence of dynamic loads. In November 1940, the suspension bridge over Puget Sound at Tacoma, Washington, U.S.A. collapsed. The indirect cause was the wind. The actual force of the wind was negligible, however, in comparison with its secondary effect, which was to start the bridge vibrating in much the same way that a strip of paper will flutter if you blow steadily along it. The rapid accelerations of thousands of tonnes of material gave rise to forces far beyond the capacity of the structure. The same cause of self-excited vibrations due to wind was partly responsible for the failure, more recently, of a number of concrete cooling towers at Ferrybridge in England.

However, these aspects of structural design are outside our province here, and we shall only deal with loads in the members of a structure arising directly from the load carried by the whole structure.

First, we must be clear about the distinction between tensile and compressive load in a member. A **tensile** load is a load which *pulls* at the ends of a member, putting it into a state of tensile stress. This is discussed in more detail in the next chapter. A typical example is the cable of a crane. A **compressive** load is opposite in sense—the member is *pushed* inwards at both ends. The supporting column of a building is a good example. The essential difference from a designer's point of view is that any shape of cross-section is satisfactory to withstand a tensile load, provided the cross-sectional area is sufficient. But a compressive load requires stiffness in addition to a minimum cross-sectional area. Obviously, a flexible steel-wire rope could not possibly sustain a compressive load. Such loads are usually resisted by steel bars having a cross-section shaped in the form of the letter L, or T, or I to give stiffness. These are called respectively, angle-sections, tee-sections and I-beams. Tubes also may be used for this purpose.

In analysing the forces in frameworks, therefore, we have to be sure not only of the magnitude of a load, but also of its nature. It is necessary, when solving the forces in a framework, to have an outline sketch of the framework itself. The nature of load within each member can then be shown as it is determined. Fig. 49 shows the convention for indicating the two types of

Fig. 49. (*a*) A tensile load: the ends of the bar pull inwards in opposition to the external outward load. (*b*) A compressive load: the ends of the bar push outwards to oppose the external inward load.

load. A tensile load is shown by arrows pointing inwards towards the centre of the bar. This represents the action of the member in resisting the external load. The compressive load is, of course, the reverse of this.

(1) Assumptions

The advantage of most framed structures is that the loads in the members are all either direct tensile or compressive loads. There is negligible bending of the members. (This advantage will be more apparent after you have studied Chapter Nine.) But this condition requires certain attention to the design of the structure. Without going into great detail, the only way to ensure that no member of a structure is subjected to bending is, first, to join all the members at their ends with frictionless pin-joints and, secondly, to ensure that when the load is applied to the structure as a whole, it is applied only at the joints, and not between them. Now it is not practicable to manufacture steel frames joined with pins at the ends, except in very special cases. In practice, members are cut to length from stock bars of material, and are jointed by bolting, riveting or welding to a gusset plate, somewhat as I have shown in Fig. 50.

The manufacturer must make some concession to the ideal joint, however.

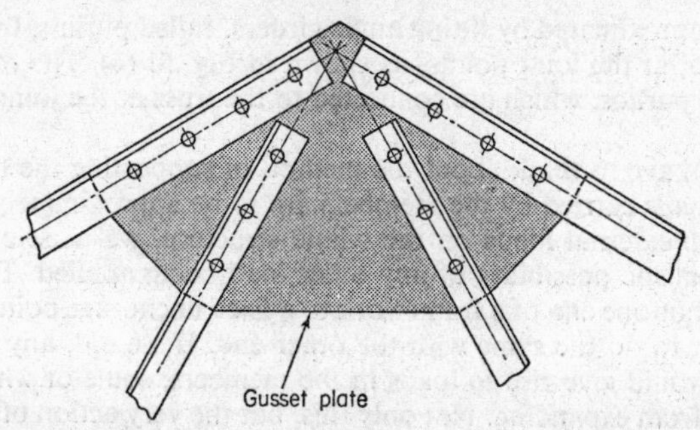

Fig. 50. A typical joint in a framework. The centre-lines of the four members should all meet at one point.

Although he cannot pin the ends, he will at least ensure that the centre-lines of all the members at a joint pass through one point. This ensures that any bending transmitted to any member is of a negligible amount.

Some practical method must also be found for applying the external load at the joints only. A large class of frameworks is that for the support of roofs. These are known as trusses. Fig. 51 shows the outline of a simple roof truss comprising nine members. Although the roof actually consists of a 'spread' load, it must, in practice, be concentrated at the five points, *A*, *B*, *C*, *D* and *E*. If, for instance, a load attachment were made midway along *AB*, as shown by the dotted arrow, it might cause excessive bending in this member. In practice,

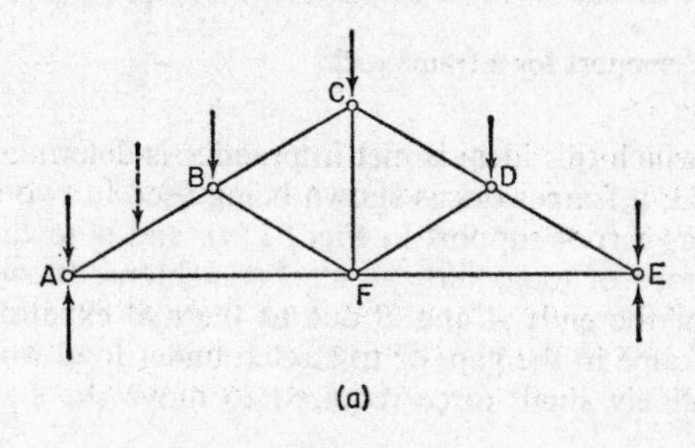

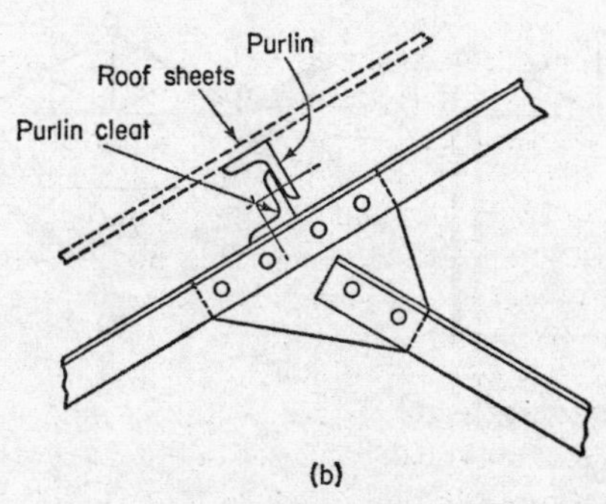

Fig. 51. A simple roof truss showing (*a*) the forces acting, and (*b*) details of a joint.

this ideal is approximated by fitting angle girders, called purlins, transversely across the roof at the joint points, as shown in Fig. 51 (*b*). The roof is supported by the purlins, which are connected to the truss at the joints by angle cleats.

Finally, we have to think about the method of supporting the framework itself. If the loads carried by the members are to be approximately as calculated from the external loads on the whole structure, we must ensure that there is no realistic possibility of any other load being applied. This means that if we anchor one end of a framework to a fixed anchorage point, we must be careful not to do the same with the other end. If we did, any change of temperature would give rise to loads in the members, some of which would be prevented from expanding. Not only this, but the very action of bolting at both ends to two fixed points might set up loads in the structure: for example, if the framework were slightly longer or shorter than the space between the anchorage points. Ideally, therefore, we should pin the framework at one point, and support it on a roller at the other, as shown in Fig. 52.

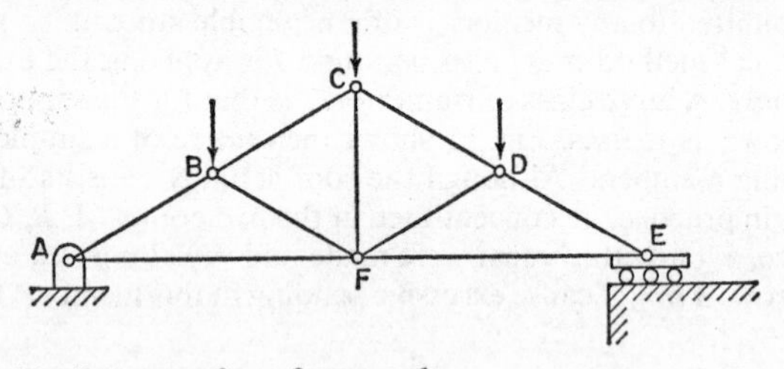

Fig. 52. The 'ideal' support for a framework.

The extent to which this ideal is met in practice is determined by the conditions. In Fig. 53, a framework is shown being used in two different situations. It is used as a roof support in Fig. 53 (*a*), and is secured at the ends *A* and *B* to the tops of long, vertical, steel stanchions. In such a case, any outward thrust of the ends *A* and *B* due to thermal expansion, to lack of initial fit of the frame in the gap, or to stretch under load would be resisted only by the relatively small force required to move the stanchion ends a

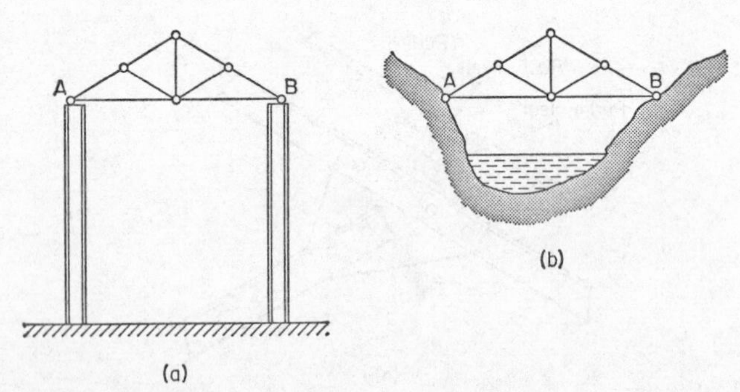

Fig. 53. The same framework used as (*a*) a roof truss and (*b*) a bridge.

short distance of perhaps a few millimetres. No harm is likely to result from securely bolting the truss at the two ends.

On the other hand, Fig. 53 (*b*) shows the same framework now used as a girder for a bridge, and attached at the ends to rock foundations. Considerable damage could result here from even a small temperature rise, or lack of fit in the gap, since the anchorage points cannot 'give' in the same way that the tops of the stanchions yield to a small load. The correct procedure in this case is for the girder to be anchored at one end only, the other end being more or less free. In a small bridge, the 'free' end may comprise a simple, flat, bearing plate on which the end may slide. In a larger structure, where a bearing plate might give rise to large frictional forces, modern procedure is to substitute a 'sandwich' plate comprising a layer of thick rubber bonded to metal plates on its upper and lower faces. Such a plate will support the heavy weight resting on it and at the same time will yield relatively easily to sideways forces. This method of support is an acceptable practical alternative to the roller assumed in the ideal structure.

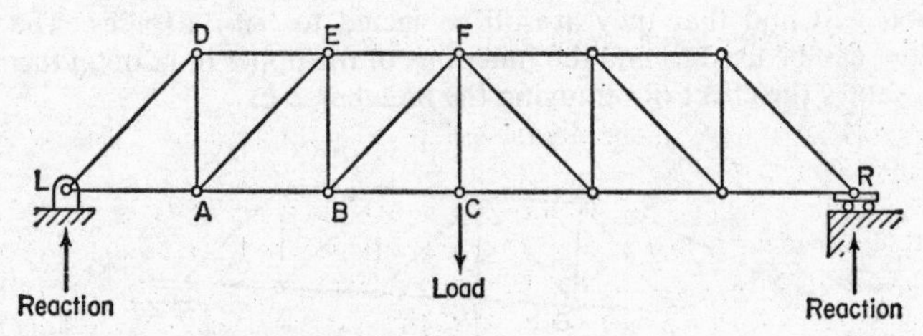

Fig. 54. A simple bridge girder with a single load.

(2) Examination of a Framework

Finding the forces in the members of a loaded structure is made simpler if we know something of the manner in which the structure is built to withstand a load. In this section we shall examine a few simple frameworks in order to find how each member plays its part in carrying the external load.

Let us first look at the simple girder shown in Fig. 54. Such a framework might be used for a small bridge. For the sake of simplicity I have shown a single load at the centre of the girder, although in practice it is probable that every lower pin-joint would carry a load. Before examining loads in the individual members of the girder, we must first make sure that we have evaluated all the external loads. The external loads in this example are the single central downward load and, secondly, the two upward reactions at the supports. I have assumed the ideal type of support, described in the previous section, with a pin-joint at the left-hand end *L* and a roller at the right-hand end *R*. Only by assuming such a support can we be certain that the two reaction forces are purely vertical and (in this case) each equal to half the central load.

For the present, we shall confine our interest to the member lettered *AB*; we shall try to find how this member helps in supporting the load on the girder. A very good way of finding out what a member is doing is to take it away and observe what happens. This is not practicable with a real framework,

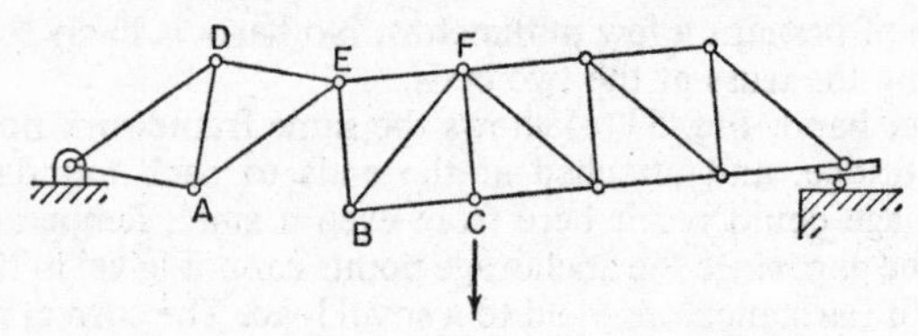

Fig. 55. When *AB* is removed, *A* and *B* move apart, indicating tension in this member.

unless it is an experimental one in a laboratory, but we can do it in imagination here. It can be seen that the removal of *AB* would result in the collapse of the girder in some such manner as I have indicated in Fig. 55.

In the partially collapsed position the points *A* and *B* have moved further apart, and to bring the girder back to its undisturbed position we should have to *pull them together again.* So the load on the girder causes the member *AB* to be pulled, i.e. to be in tension.

If you adopt this argument for any other of the six lower horizontal members, you will find that they are all subjected to tensile forces. The same procedure can be used to find the functions of the upper horizontal members. Fig. 56 shows the effect of removing the member *DE*.

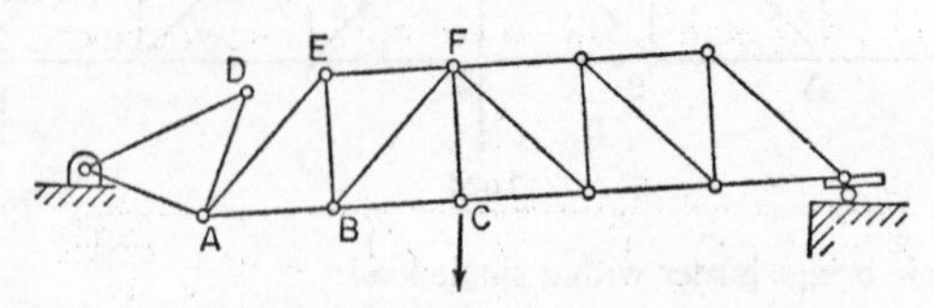

Fig. 56. Removal of *DE* causes *D* and *E* to move together, indicating compression in this member.

Here you can see that *D* and *E* are brought closer together as the girder collapses, and would have to be pushed apart to cancel the damage. The member *DE* must, then, be subjected to a compressive load. A similar analysis would show all four of the upper horizontal members to be in compression.

The function of the inclined members is not quite so easy to appreciate but let us try the effect of removing *AE*. We see in Fig. 57 that the original square *ABDE* is distorted to a rhombus, and that *A* and *E* have been brought closer together. The same result applies to all six inclined members, which are thus in a state of compression due to the load on the girder.

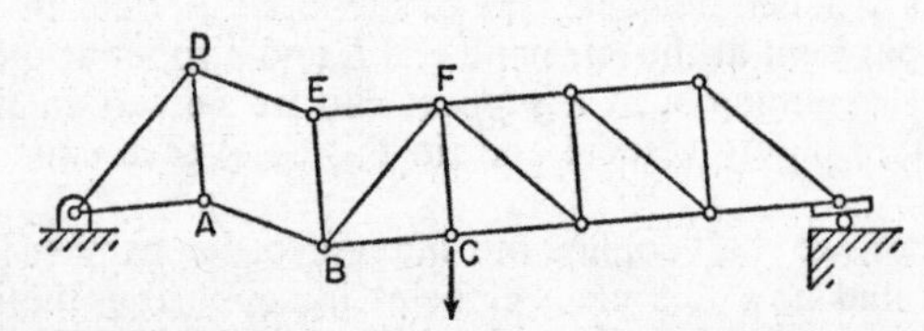

Fig. 57. Removal of *AE* causes *A* and *E* to move closer together, indicating compression in this member.

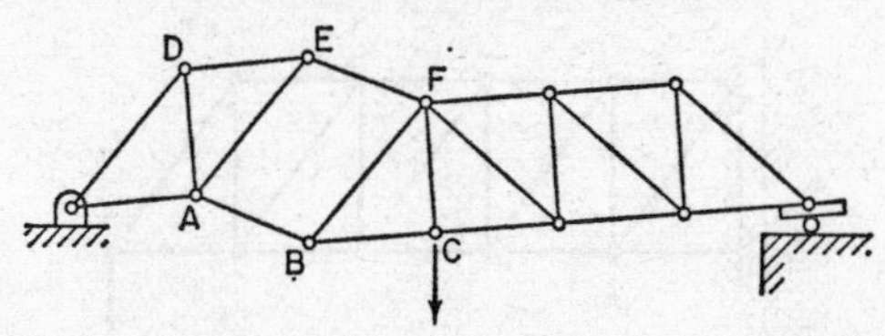

Fig. 58. Removal of *BE* causes *B* and *E* to move apart, indicating tension in this member.

Lastly, Fig. 58 shows the effect of removing a typical vertical member such as *BE*. Points *B* and *E* move apart during collapse, indicating tension in this member. It can be similarly shown that all five vertical members are in a state of tension.

We must not generalize from these results. It might be tempting to assume that the upper and lower horizontal members of a girder are always respectively in compression and tension, and that the inclined members are always in compression. A few further examples will show that no such general rules can be formulated. Suppose, for instance, we turn our previous example upside-down and apply the load at the top, as shown in Fig. 59. This is quite a common method of construction for small bridges. You can easily show for yourself that in this example the lower horizontal members, such as *DE*,

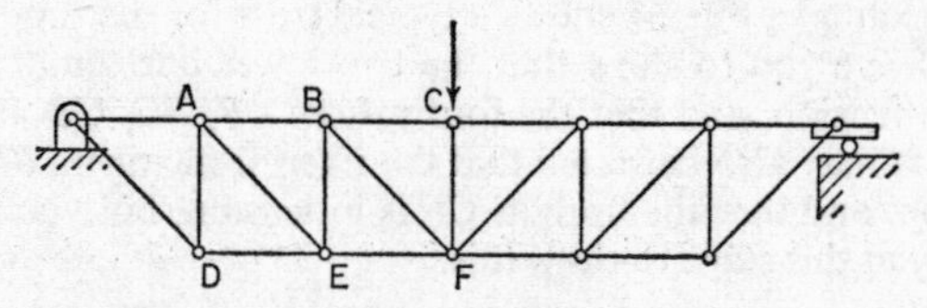

Fig. 59. The girder of Fig. 54 turned upside-down.

would again be in tension, and the upper ones, such as *AB*, would be in compression. But you would also find that the sloping members are now all in tension, and the uprights are in compression. You could have worked this out in a simpler manner if you had observed (as you may have done) that the girder of this second example is identical in all respects to that of Fig. 54 except that the *directions of the external loads are reversed.* It is a reasonable assumption (and also a correct one) that reversing the directions of all the external loads on a simple pinned structure will reverse the direction of load in every member of that structure.

Now let us examine a different type of framework. Fig. 60 shows a single load carried at one end of a horizontal girder whose other end is fixed to a wall. I have had to assume only one of the two support points (*P* in this example) to be a pin-joint, the other support being a roller. Such a girder is called a **cantilever.** Two possible arrangements are shown, and you can see that the girder is identical in both cases but is turned upside-down in Fig. 60 (*b*). Using the technique of removing the member you are analysing, you should be able to deduce that, in both cases, the upper horizontal members are now in tension and the lower ones are in compression. The sloping members will be in tension in Fig. 60 (*a*) but in compression in Fig. 60 (*b*).

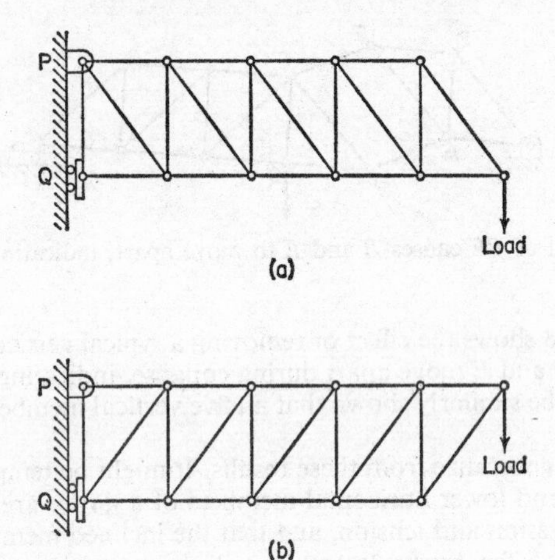

Fig. 60. Two examples of a cantilever framework.

Conversely, the upright members will be in compression in Fig. 60 (*a*) and in tension in Fig. 60 (*b*).

For a final example, Fig. 61 shows a typical truss for the support of a roof. I shall again leave you to show that the two lower horizontal members *AF* and *FE* are in tension, and that the four rafters *AB*, *BC*, *CD* and *DE* are in compression. It may also be stated that the interior members *BF* and *DF* are in compression, and that the upright *CF* is in tension, but you may not find it quite so easy at this stage to show this.

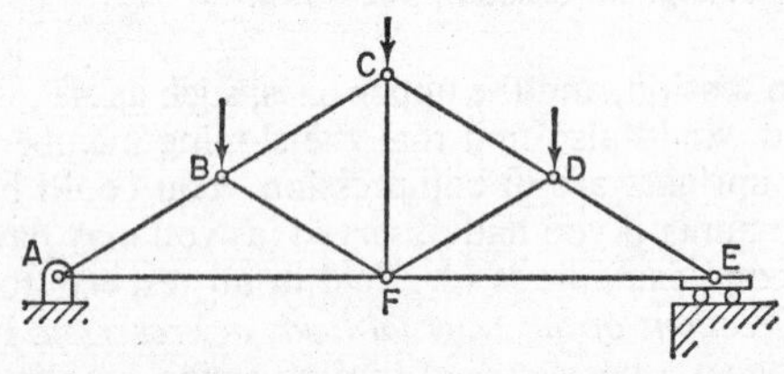

Fig. 61. A simple roof truss.

(3) Determination of the Load

We can now develop the argument of the previous section to determine the actual value of the load in a member, in addition to its function in the framework. Suppose we go back to the simple bridge girder of Fig. 54. This is reproduced in Fig. 62, where the frame is in the process of collapsing because we have removed the member *AB*. Let us bring the frame back to the uncollapsed position by applying a pull at the two points *A* and *B*.

We shall have to apply a pull which is equal in magnitude to the pull which was being exerted by the member we have removed. Call this force *f*. In Fig. 62 (*b*) I have shown the first two panels of the girder, with this force *f* in place of the member AB. To make a numerical example, I have given dimensions to the structure, and a value of 2 kilonewtons to the load. The reaction forces at the support points will be each 1 kilonewton. If the force f were not there, the left-hand upward reaction force of 1 kilonewton would cause the two left-hand panels to turn in a clockwise direction about the point E as a hinge. Indeed, this is what is happening when the frame collapses.

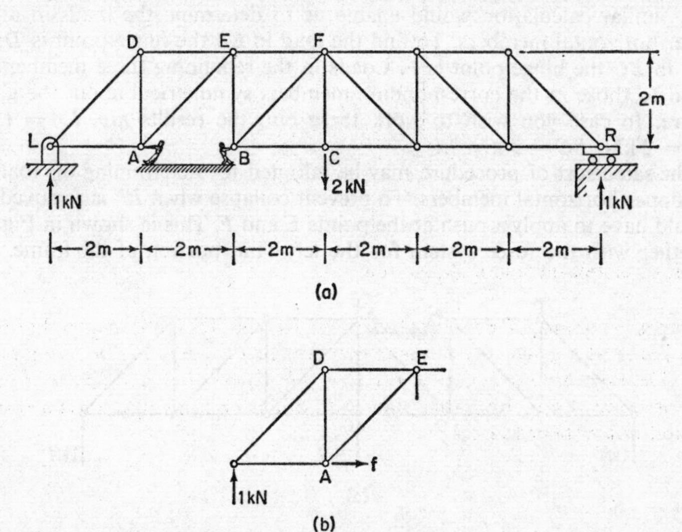

Fig. 62. AB is removed. The frame is held together by pulls at A and B equal to those exerted by the missing member.

So the reaction force has a moment about the point E. Collapse can be prevented if the force f has an equal and opposite moment about the same point. We can thus find f by writing an equation of moments about E:

$$(1 \times 4) - (f \times 2) = 0$$

giving a value of 2 kilonewtons for the tensile force f in the member AB.

We can also obtain the same result by considering the hinging of the right-hand section of the girder about E. In this case, there are two external loads: the central load of 2 kilonewtons and the right-hand upward reaction. Fig. 63

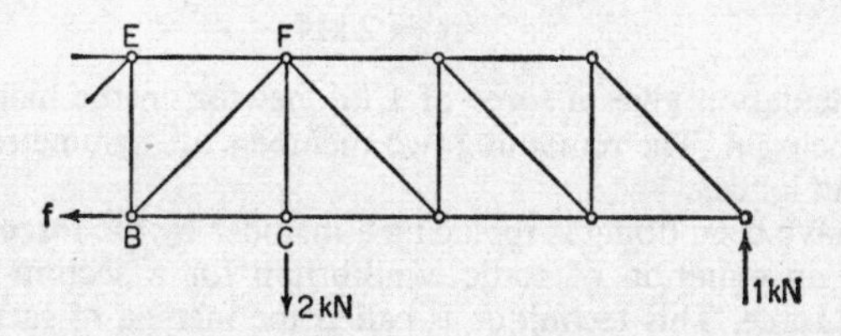

Fig. 63. Equilibrium of the right-hand portion of the frame in Fig. 62; f is the force in the missing member AB.

shows the situation. The equation of moments for equilibrium about the same point E is now:

$$(2 \times 2) - (1 \times 8) + (f \times 2) = 0$$

giving $f = 2$ kilonewtons as before, although the calculation is very slightly longer.

A similar calculation would enable us to determine the loads in all six lower horizontal members. To find the load in LA the hinge-point is D; for that in BC the hinge-point is F. Loads in the remaining three members are equal to those in the corresponding members symmetrical about the girder centre. In case you wish to work these out, the results are, $LA = 1$ kN, $AB = 2$ kN, $BC = 3$ kN.

The same sort of procedure may be adopted for determining the loads in the upper horizontal members. To prevent collapse when EF is removed, we should have to apply a push at the points E and F. This is shown in Fig. 64, together with the force system for the left-hand portion of the frame.

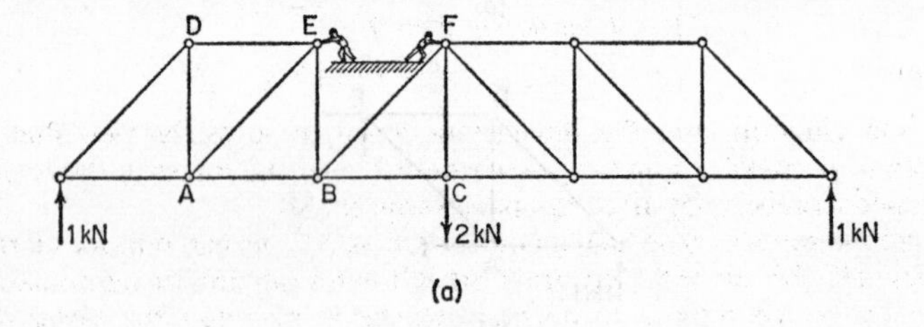

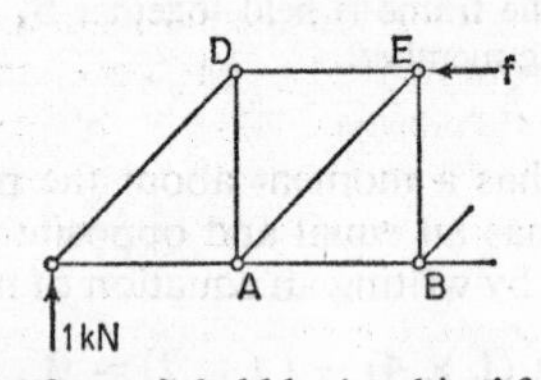

Fig. 64. *EF* is removed. The frame is held by 'pushing' forces at *E* and *F* equal to the forces exerted by the missing member.

Here, f is evaluated by writing an equation of moments about the point B:

$$(1 \times 4) - (f \times 2) = 0$$

Therefore $$f = 2 \text{ kN}$$

A similar calculation gives a force of 1 kilonewton in the member DE, the hinge-point being A. The remaining two members are symmetrical and carry corresponding loads.

What we have been doing is replacing a member by the force it exerts, and then writing an equation of static equilibrium for a section of the frame carrying this force. This technique is called the **method of sections.** We can extend the method to find the forces in the sloping members. Suppose we now remove all three members AB, AE and DE, and replace them by correspond-

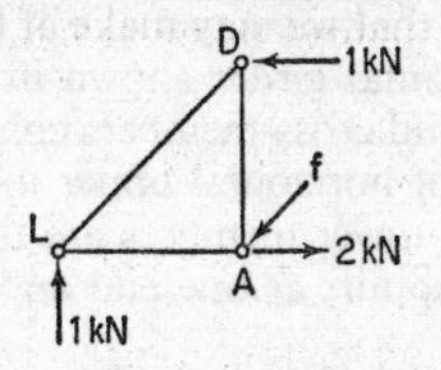

Fig. 65. Equilibrium of a portion of the frame; f is the unknown force in the member AE.

ing forces, since we know the forces in AB and DE. Fig. 65 is a diagram of the forces acting on the left-hand portion of the frame remaining.

We again call the required force f, i.e. the force in the sloping member AE. We have already shown it to be compressive, and I have shown it so on the diagram. We could write an equation of moments about the point L or about the point D, but it is easier to equate the forces in any one direction to zero. For example, the total force acting to the left is zero:

$$1 + f \cos 45° - 2 = 0$$

Therefore

$$f = 1{\cdot}414 \text{ kN}$$

(From the diagram, and the dimensions given, it is easily seen that the members slope at 45°.) You can show similarly that the forces in the remaining inclined members are all of the same value as this.

To find the force in a vertical member such as BE, we can remove all three members AB, BE and EF, and draw the left-hand portion of the remaining framework (or the right-hand portion, whichever promises the easier solution). Observe that one of the members removed must be that in which the force is required. Forces are added in place of the missing members, and an equation of equilibrium obtained. Fig. 66 shows the force diagram when members AB, BE and EF are removed, and you will see again that AB and EF have been replaced by the known values of the forces in these members, both as to magnitude and direction. If we had chosen the right-hand portion of the frame instead, the forces would have had the same values (2 kilonewtons each, in this case) but would have been in the opposite directions.

The simplest equation here is that of vertical equilibrium, from which it can be easily seen, without actually writing the equation, that f is 1 kilonewton. The same result would be obtained for all five vertical members.

(4) Some Conclusions

I have already warned against making generalizations concerning the functions of various components of a pin-jointed framework. There are,

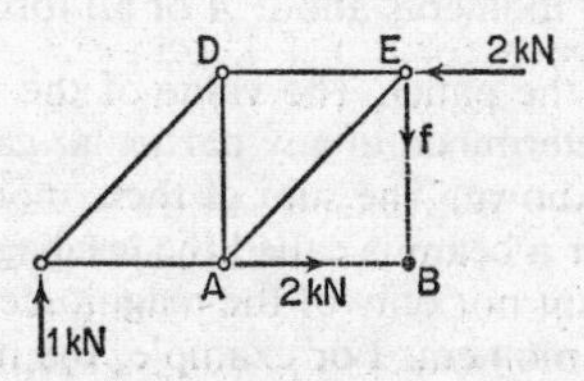

Fig. 66. Diagram to determine the force in member BE.

however, certain deductions that we may make at this stage. For this purpose we shall consider the horizontal girder shown in Fig. 67. It has horizontal upper and lower members, and cross-members consisting of both vertical and inclined members. The set of horizontal upper members is called the upper chord of the frame, and the lower members are the lower chord. The girder is assumed to have a pin-support at one end and a roller at the other. For

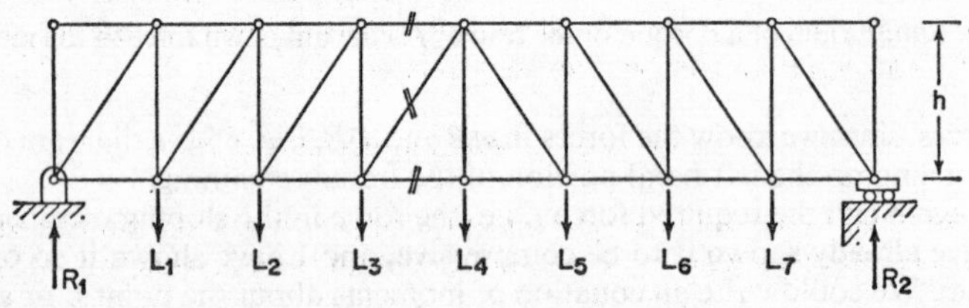

Fig. 67. A loaded N-girder.

fairly obvious reasons, such a framework is called an N-girder. We shall assume that it carries loads at every intermediate joint along the lower chord.

Suppose we wish to determine the loads in the three marked members in the fourth panel from the left in Fig. 67. We can remove these three members and replace them by the three forces they exert. We will call these forces f_U, f_L and f_I, denoting respectively the forces in the upper, lower and inclined member. From the considerations of Section 2 of this chapter, we know that f_U will be compressive and f_L will be tensile. We can now draw the left-hand three panels, with all the external loads, and with the three forces designated above.

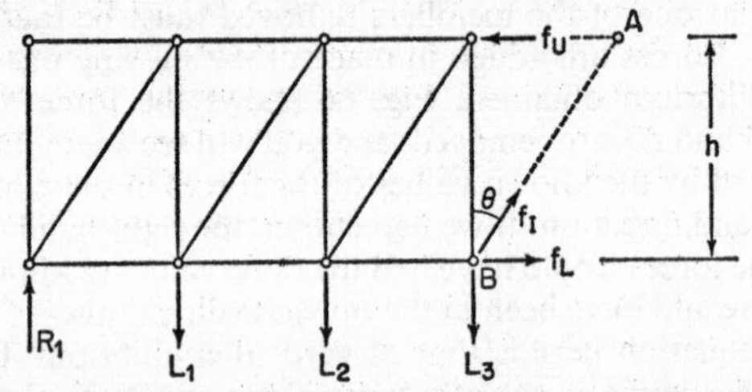

Fig. 68. Equilibrium of the left-hand three panels of the girder shown in Fig. 67.

From the work on the first example of the previous section, we know that f_L may be found by writing an equation of moments about the hinge-point *A*. We can write this equation in very general terms thus:

$$f_L \times h = \text{sum of moments about } A \text{ of all forces to left of } A$$

where *h* is the height of the panel. The value of the right-hand side of this equation can easily be determined in any particular case where the values of loads and reactions are known. The sum of these moments about any point (such as *A*) on a girder or a beam is called the **bending moment**. The bending moment must take account not only of the magnitude, but also of the direction of each component moment. For example, the moment of R_1 is in the opposite direction to the moments of the loads L_1, L_2 and L_3.

The determination of the bending moment at various points along a beam or girder is an essential procedure in design, and we shall have something more to say on this subject when we deal with the bending of beams in Chapter Nine. The variation of bending moment depends on the loads on a beam, the manner in which they are distributed, and the way the beam is supported.

The important conclusion to note at this stage is that the load in a member of the lower chord is proportional to the bending moment at that point on the girder. The equation can be rewritten thus:

$$f_L = \frac{1}{h} \times \text{bending moment at } A$$

We can obtain the load in a member of the upper chord by a similar equation of moment about the point *B*. We should obtain the equation:

$$f_U = \frac{1}{h} \times \text{bending moment at } B$$

We can therefore regard the members in the upper and lower chords of a girder as resisting the bending moment at that particular section of the beam. The greater the bending moment, the greater the load in these members. If you have already read Chapter Nine, you may be able to see an affinity between the upper and lower flanges of an I-beam and the members of the upper and lower chords of a girder.

For a girder which is simply supported at the two ends and carries a load in between, the bending moment is zero at the ends and has its maximum value near the centre—exactly at the centre if the load is distributed uniformly from end to end. If the girder is of the same height throughout, as in our example, it follows that the cross-sectional area of the members in the upper and lower chords will have to increase as the bending moment increases, i.e. towards the centre of the girder. The chord members close to the centre would have to be large to carry the high loads, while the outer ones would need to sustain only very light loads. But if we look at the expression for the value of f_L or f_U, we can see that the value of the load may remain the same if we increase the height h of the panel in proportion to the bending moment at that point. This generally makes for a much more pleasing and satisfactory design for large and medium-sized girders. Our original girder of constant depth, so modified, might look more like Fig. 69.

Such a design makes for economy of material in addition to being more attractive. For a span of 350 feet, such as the girders on the Forth railway bridge, each incorporating 872 tons of steel, economy is a factor not to be

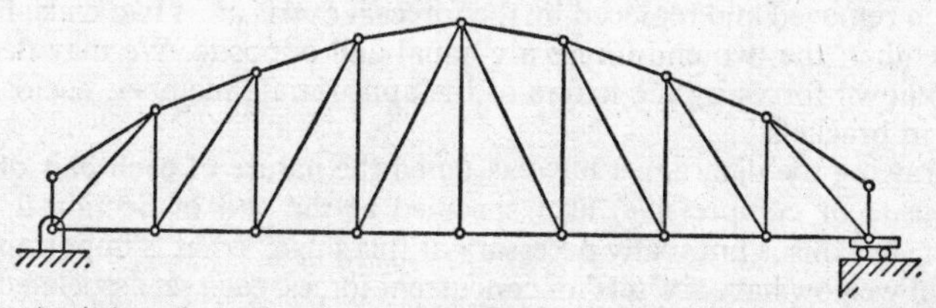

Fig. 69. A girder of varying depth. The depth increases as the bending moment increases.

lightly dismissed. On a still larger scale, the central span on the Quebec bridge over the St. Lawrence river is 640 feet long and the girder contains over 5000 tons of steel. It may be noted in passing that economy of material in this connection does not always result in the least outlay. The construction of the Quebec bridge required that the central span be hoisted up complete from pontoons from the river below. On two occasions during this process, in 1907 and in 1916, the whole span collapsed and fell into the river. It must be added that this double disaster had no connection with the feature of design we are discussing.

If we now turn our attention to the load f_1 in the inclined member of Fig. 68, we may write an equation of vertical equilibrium as follows:

$$f_1 \cos \theta = L_1 + L_2 + L_3 - R_1$$

where θ is the angle of inclination of the member to the vertical. In more general terms:

$$f_1 \cos \theta = \text{algebraic sum of vertical external forces to left of section}$$

The quantity on the right-hand side of this equation again depends on the loads, and their manner of distribution. This quantity is called the **shear force** at the particular point along the beam or girder. For a girder having loads distributed equally along it, and simply supported at the ends, the shear force would have a maximum value at the two ends, and would be zero in the middle. Thus, the inclined members of a girder of this type can be thought of as offering resistance to the shear force at the appropriate point along the girder.

(5) Force Polygons and the Maxwell Diagram

We have seen earlier in this chapter that a load within a member can be deduced, using a simple equation of equilibrium, by removing that member from the framework and replacing it by the force it is assumed to exert on the framework. There is no limit to the number of members which may be removed; for instance, Fig. 68 shows three members removed. We may even remove *all* the members of a framework and replace them by forces. Let us do this with the simple roof truss of Fig. 52.

The truss is redrawn in Fig. 70, with specific values of loads inserted at the three upper joints. The values of the reaction forces at the two ends have been calculated from a simple equation of moment of all external forces about one end or the other, exactly as we calculated beam reactions in Chapter Three. Below the loaded framework I have shown the situation when every member has been removed and replaced by the forces it exerts at its two ends. For any one member, the two end-forces are equal and opposite. We may designate the unknown forces by the letters of the appropriate member, enclosing the letters in brackets.

In drawing the diagram, I have assumed the nature of each pair of forces (i.e. tensile or compressive) as I specified at the end of Section 2 of this chapter, but this is not really necessary at this stage. What is important is the fact that we now have six 'sets' of concurrent forces, each set associated with a particular joint of the framework. This situation invites a solution using the **polygon of forces.** We know from Chapter Three that, for a number of forces

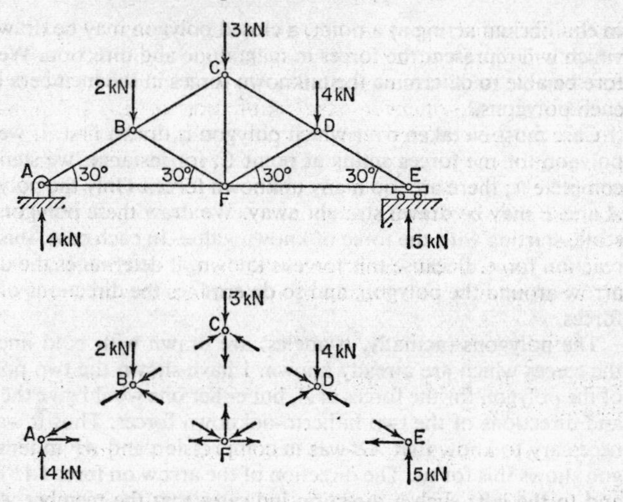

Fig. 70. When all members of a frame are replaced by forces, we have a set of concurrent forces for each joint.

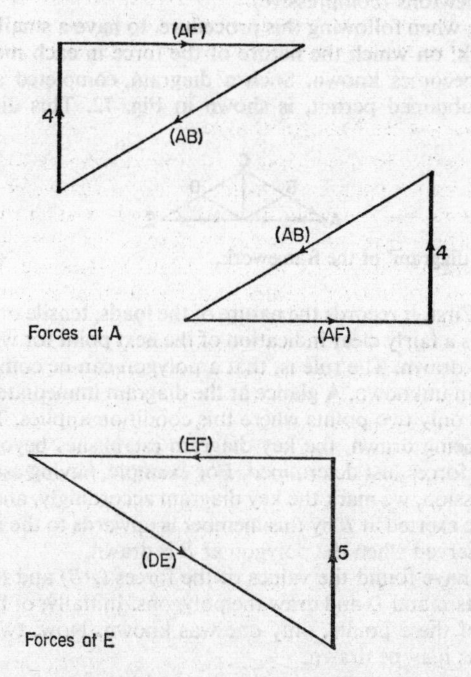

Fig. 71. Polygons for the forces at *A* and *E*. Bold lines represent the forces which are known.

in equilibrium acting at a point, a closed polygon may be drawn, the sides of which will represent the forces in magnitude and direction. We should therefore be able to determine the unknown forces in the members by drawing six such polygons.

Care must be taken over which polygon is drawn first. If we try to draw a polygon for the forces acting at point *C*, for instance, we shall be unable to complete it; there are too many unknown forces. Only the polygons at points *A* and *E* may be drawn straight away. We draw these polygons to a suitable scale, starting with the force of known value. In each case, this is the vertical reaction force. Because this force is known, it determines the direction of the arrow around the polygon, and so determines the directions of the other two forces.

The polygons (actually, triangles) are drawn with bold lines to represent the forces which are already known. I have shown the two possible versions of the polygon for the forces at *A*, but either one would give the correct values and directions of the two hitherto-unknown forces. Thus it was not actually necessary to know that *AB* was in compression and *AF* in tension; the polygon shows this for us. The direction of the arrow on force (*AB*) is downwards and to the left; such a direction indicates that the member *AB* pushes outwards at the point *A*, and hence must be in compression. Measurement or calculation from the triangle tells us that (*AF*) is 6·93 kilonewtons (tensile), (*AB*) is 8 kilonewtons (compressive), (*EF*) is 8·66 kilonewtons (tensile) and (*DE*) is 10 kilonewtons (compressive).

It is necessary, when following this procedure, to have a small key diagram of the framework, on which the nature of the force in each member can be indicated as it becomes known. Such a diagram, completed so far as the results already obtained permit, is shown in Fig. 72. This diagram serves

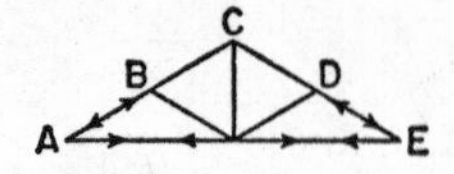

Fig. 72. The 'key diagram' of the framework.

three purposes. First, it records the nature of the loads, tensile or compressive. Secondly, it gives a fairly clear indication of the next point for which the force polygon is to be drawn. The rule is, that a polygon can be completed if only two forces remain unknown. A glance at the diagram immediately shows that *B* and *D* are the only two points where this condition applies. Thirdly, when the polygon is being drawn, the key diagram establishes beyond doubt the *directions* of the forces just determined. For example, having established that *AB* is in compression, we mark the key diagram accordingly, and we can then see that the force exerted at *B* by this member is upwards to the right, and this fact must be observed when the polygon at *B* is drawn.

Now that we have found the values of the forces (*AB*) and (*DE*), we proceed to the points *B* and *D* and draw the polygons. Initially, of the four forces acting at each of these points, only one was known. Now, two are known, and the polygons may be drawn.

Attention must be given to the order in which the four forces are drawn. We must start with the two forces whose values are known. At *B*, these are the downward load of 2 kilonewtons and the force of 8 kilonewtons in *AB*. The two remaining forces (*BF*) and (*BC*) may then be drawn, the sides of the polygon being parallel to the appropriate member. Within these limits, there are alternatives; for instance, (*AB*) may be drawn after the 2 kN load instead of before it. But the magnitude and direction of the two remaining forces (*BF*) and (*BC*) will be found to be the same no matter how the polygon is drawn.

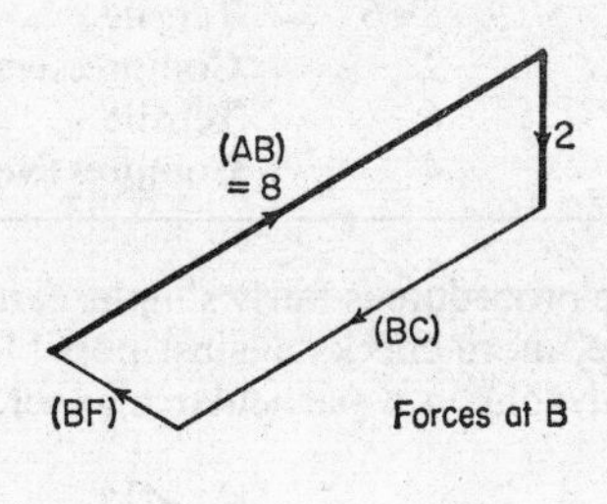

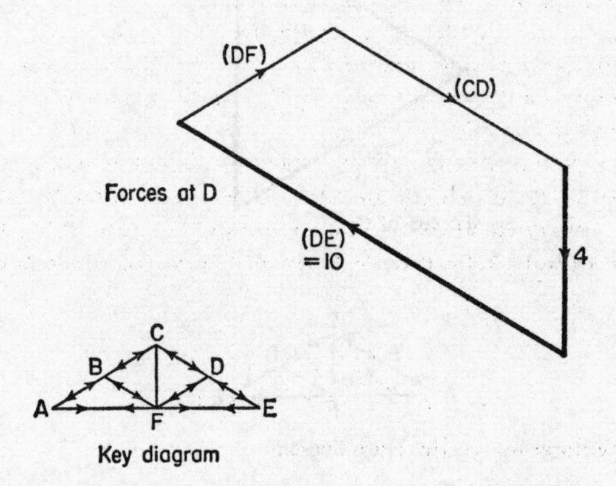

Fig. 73. Force polygons for points *B* and *D*, and the key diagram.

The direction of (*BF*), upwards and to the left, establishes this force as compressive when marked on the key diagram; likewise, (*BC*) acting downwards to the left indicates again a compressive load in the member. (*BF*) will be found to have a value of 2 kilonewtons and (*BC*) 6 kilonewtons. The polygon at *D*, similar in shape, gives (*DF*) as 4 kilonewtons and (*CD*) as 6 kilonewtons.

The key diagram now indicates only one further force—that in member *CF*—to be found. For this, we may draw either the polygon at *C* or that at *F*; it is not necessary to draw both, although one could do so as a check. Fig. 74 shows the polygon for *C*. At this final stage, three of the four forces are known; if they are correct, the polygon should indicate that force (*CF*) is vertical. You can see that it is, and that its value is 6 kilonewtons. We can now draw up a table listing the value and nature of all the loads in all the members.

Member	*Load, kN*	*Nature*	*Symbol*
AB	8	Compressive	←—→
BC	6	Compressive	←—→
CD	6	Compressive	←—→
DE	10	Compressive	←—→
AF	6·93	Tensile	→—←
EF	8·66	Tensile	→—←
BF	2	Compressive	←—→
CF	6	Tensile	→—←
DF	4	Compressive	←—→

Although the above procedure is fairly simple, certain work can be avoided and, at the same time, more checks against possible error can be made, by drawing the force polygons in a particular manner. You may have already

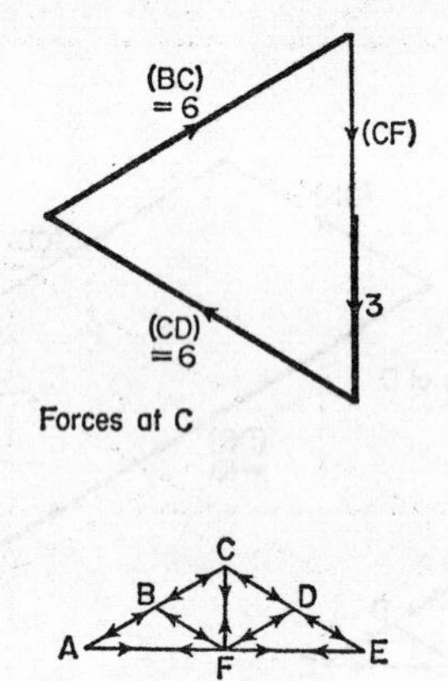

Fig. 74. Polygon for forces at *C*, and the final key diagram.

noticed that the vector result of one diagram is used to help in the drawing of another. This duplication can be avoided by combining the equal and opposite forces in a member into one vector, using what is called a **Maxwell diagram.**

In order that we may do this, we have to avoid putting arrows on to the vectors, which are to be capable of being read in either direction. Secondly, the force in a given member has to be designated in a special manner, which has come to be known as **Bow's notation.** The principle of this notation is that, instead of lettering the joints at the ends of a member, the member is designated by letters representing the *spaces* on either side of it. In Fig. 75 I have shown the outline of the previous framework, lettered according to Bow's notation (although you will note that letters have been used only in the spaces outside the framework, numbers being used for the spaces in between members). Below the framework diagram, in Fig. 75 (*b*), I have shown the completed Maxwell diagram.

The first point to note is that sufficient spaces must be lettered on the frame diagram to ensure that every *force* can be designated by a different pair of letters, or numbers, on either side of it. Thus the downward load of 3 kilonewtons at the apex of the truss is designated *c–d*; the left-hand upward reaction of 4 kilonewtons is *a–b*, and so on. Proceeding more or less as in the previous analysis, we now draw a polygon for the left-hand joint, which was joint *A*, but is now joint *a–b–1*. We must faithfully observe one rule throughout: we must draw the force vectors in order, taken in the same direction around the joint. We shall arbitrarily take the direction to be clockwise.

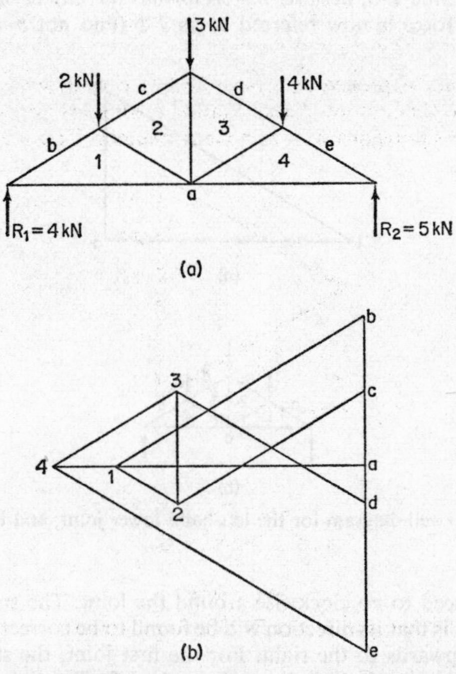

Fig. 75. (*a*) Outline of the framework from Fig. 70, lettered according to Bow's notation. (*b*) The Maxwell diagram.

Drawing the upward vertical force of 4 kilonewtons first (because we know the value), this is denoted *a–b*. This gives the first vector, *a–b* on the Maxwell diagram. The order of lettering on the vector follows the direction of the force—in this instance, upwards.

Proceeding clockwise, the next force we encounter is that in the member *b–1*; so a line is drawn from *b* on the Maxwell diagram parallel to the member *b–1*. A horizontal line through *a* represents the third force, that in member *1–a*, and intersects the second line at point *1*. The Maxwell diagram so far is shown in Fig. 76 (*a*).

At this stage we must indicate on the key diagram the directions of the forces for the two members we have solved. We refrain from putting arrows on the Maxwell diagram, but we know that the force *a–b* is upwards, and so we read around the triangle in the direction consistent with this. This gives *b–1* downwards to the left, and *1–a* horizontal to the right. We may put corresponding arrows on the key diagram, on the appropriate members, at the same time inserting arrows in the opposite directions at the other ends of the members. When this is done, as in Fig. 76 (*b*), the members *b–1* and *1–a* can be seen to be respectively in compression and tension.

We are now able to proceed to the next joint, that where the 2 kN load is applied. Our choice is governed, as before, by the fact that the force *b–1* has now been determined, leaving only two unknown forces (*c–2* and *2–1*). We start with the force *1-b*, *because the vector has already been drawn for this.* Note that this force is now referred to as *1–b* (and not *b–1* as previously)

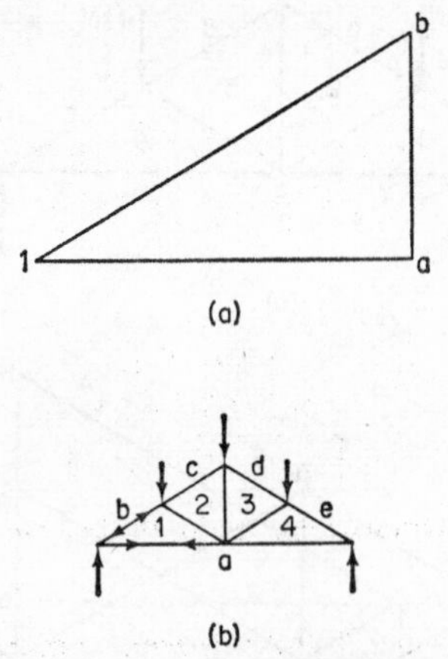

Fig. 76. The Maxwell diagram for the left-hand lower joint, and the corresponding key diagram.

because we agreed to go clockwise around the joint. The special advantage of this notation is that its direction will be found to be correct on the Maxwell diagram, i.e. upwards to the right. For the first joint, the same line on the Maxwell diagram was *b–1*, downwards to the left. Force *1–b* is followed in turn by *b–c*, which is known to be 2 kilonewtons vertically downwards. The direction of *c–2* enables us to draw a line through *c*, and that of *2–1* enables us to draw the line through *1* to intersect it. The Maxwell diagram now looks like Fig. 77.

Establishing the direction around the polygon by the known direction of any one of the four forces, for example *b–c*, we obtain the direction of force *c–2* as downwards to the left, and that of *2–1* as upwards to the left. Marking these on the key diagram, with opposite-pointing arrows at the other ends, indicates compression in both these members.

Sufficient has been stated to make the derivation of the Maxwell diagram fairly clear, I hope. Referring back to Fig. 75 (*b*), having established the vector line *c–2*, we can extend the polygon for the joint at the apex *2–c–d–3*. Starting with the existing line *2–c*, we draw *c–d* downwards, making its length *3* units. Next we fix point *3* by the intersection of lines through *d* and *2*, parallel respectively to the forces *d–3* and *3–2*. We then move on to the next joint, *3–d–e–4*, which finally fixes the point *4*.

This completes the Maxwell diagram, and from it may be scaled the magnitudes of the forces in all members of the framework. We have not specifically drawn the force polygon for the central lower joint, or for the right-hand support joint, but you will see that the complete diagram includes these polygons, which are drawn automatically, so to speak. The polygon for the central lower joint, for instance, is *a–1–2–3–4*. The point *4* is the last one to be drawn on the diagram, and we determine it from the intersection of the

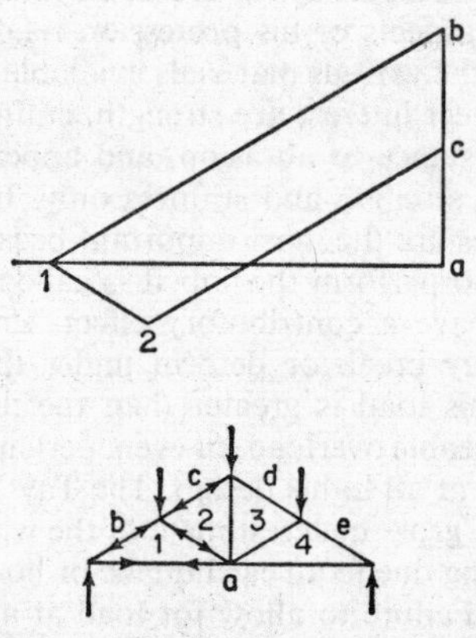

Fig. 77. The Maxwell diagram completed for the first two joints, and the corresponding key diagram.

oblique lines through *e* and *3*. But it is seen from the frame diagram that the force *a–4* must be horizontal. If the diagram has been accurately drawn, line *a–4* will be found to be horizontal. This forms a useful check on the correctness and accuracy of the drawing. Compared with the technique of constructing separate force polygons for each joint, the Maxwell diagram requires appreciably less drawing.

One final point should be noted. The five letters *a* to *e* on the Maxwell diagram form in themselves a separate force 'polygon', representing the external forces on the framework as a whole. This is why it is good practice to letter the external spaces distinct from the internal ones. The 'polygon' in this case is a straight line, with all the forces vertical. When you become practised in the technique of drawing these diagrams, you will find it easier and quicker to draw this 'external-force' polygon complete, before tackling the forces in the component members of the framework.

CHAPTER EIGHT

BEHAVIOUR OF MATERIALS: STRESS ANALYSIS

Since an engineer is almost always concerned with making something, one of the most important aspects of his profession relates to the nature, properties and diversity of the various materials available to him. The properties that are of the most direct interest are strength, stiffness, workability, resistance to corrosion, resistance to abrasion, and appearance. Applied mechanics is concerned with strength and stiffness only. In a sense, one can say that these two properties are the most important because they determine the ability of the material to perform the job it is called upon to do, although other properties may have a contributory effect. Engineering components usually fail because they break or deform under the load imposed upon them, either because this load is greater than the designer anticipated, or because of some unforseeable overload, or even, perhaps, because the designer failed to allow for load at all in his design. The Tay Bridge disaster of 1879 was due principally to a gross underestimate of the wind force on the bridge. The collapse of a building due to an earthquake or bomb is an example of an unforseeable overload. Failure to allow for load at all is not uncommon in the design and manufacture of many cheap kitchen gadgets and children's toys.

An essential requirement of any engineering material is that it should behave satisfactorily under load. This does not mean merely that the component must not break: it must not deform beyond a certain acceptable limit. It would be quite possible to design and erect a perfectly *safe* road bridge of strong rubber stretched between the abutments, but it would be a matter of considerable difficulty, and understandably so, to induce any driver to make use of it.

The examination of these properties of strength and stiffness is called **stress analysis**. This has superseded the earlier, and perhaps more familiar name of 'strength of materials'. Stress analysis examines the theory underlying the behaviour of materials under all conditions of load, and establishes rules and formulae capable of deciding such diverse questions as the sizes of the various components of a steel bridge, the thickness of columns and walls of a modern multi-storey building, the movement of a skyscraper due to the force of the wind, the wall thickness of a steam boiler, the diameter of the propeller shafts of a large liner, and a host of others. In this book, however, we shall only be able to examine the very simplest aspects of stress analysis.

(1) Tensile Stress and Compressive Stress

Suppose we take a bar of a certain material, steel for example, of uniform cross-sectional area, and pull the ends of it apart in a machine until it breaks in two. (This, incidentally, is one of the more interesting experiments undertaken by engineering students.) We can measure the load, and we can also measure the area of cross-section of the bar. For instance, we might take a bar having a square cross-section, of length of side 1 centimetre, and we might find that a force of 45 kilonewtons was required to break it in two. Then we might reasonably assume that a similar bar of the same material, but having a square cross-section of 2 centimetres length of side, having four times the cross-sectional area, would require four times the load to break it, i.e. 180 kilonewtons. And using the same line of reasoning, a bar having a circular cross-section of diameter 4 millimetres would have a cross-sectional area of $\pi/4 \times (0{\cdot}4)^2 = 0{\cdot}126$ square centimetres, and should therefore withstand a load of $(45 \times 0{\cdot}126) = 5{\cdot}65$ kilonewtons. It is fair to add at this stage that, providing certain precautions were observed, the reasonable assumptions we have made would be supported by direct experimental observation. The necessary precautions will be discussed later in this chapter.

So we could take 45 kilonewtons, the load withstood by 1 square centimetre, as a sort of standard for the material, and use this as a factor for calculating the load on any fraction or multiple of this area. We are beginning to generalize our observations, and to *predict* the behaviour of as yet unused bars of material. The load required to break one square centimetre of the material is called the **ultimate stress** for that material, and you can see that it represents a property of the material, steel in this case, rather than a property of any particular bar. Although I have given a value of 45 kilonewtons per square centimetre, the stress can be stated in any other units we care to nominate, such as tons per square inch, pounds per square inch, or newtons per square metre. This last is consistent with the MKS system, and the value is 450×10^6 newtons per square metre, or more simply, 450 meganewtons per square metre. We can now infer that *any* steel bar having a uniform cross-section will break when the load in it is such as to cause a stress, or load per square centimetre, of 45 kilonewtons. We thus have a method of calculating the breaking load of *any* steel bar, providing the mode of loading is the same, i.e. a direct tensile (pulling) load.

Suppose, then, we are asked to provide a steel-wire rope for a crane, capable of lifting a total mass of 8 tonnes. The weight of 8 tonnes is $8 \times 1000 \times 9{\cdot}81 = 78{\cdot}5$ kilonewtons. Since we know that one square centimetre will take 45 kilonewtons, we shall require a cross-sectional area of steel of $78{\cdot}5/45 = 1{\cdot}75$ square centimetres. (We must remember that this is the area of *steel* required, not of rope cross-section: a wire-rope cross-section contains a high proportion of space between the strands.)

(2) Factor of Safety

I hope that most readers will have spotted a grave weakness in the preceding argument. What we have done is to provide *just enough* steel to support the load; in fact, the load it is to support will, by our reckoning, be just sufficient to break it. This is manifestly unsatisfactory. If we are going to apply load to a material, we must keep the load down to an acceptable maximum limit, to ensure no reasonable possibility of breakage. In the present example, it would be a most unwise risk to permit a stress in a crane cable which was greater than about one-third of the breaking stress. With this proviso in mind, we should really specify a minimum cross-sectional area of three times our previous value, i.e. 5·25 square centimetres. With the same maximum load, the corresponding stress (i.e. load per square centimetre of cross-section) will be $\frac{1}{3} \times 45 = 15$ kilonewtons per square centimetre.

The highest stress (i.e. load per unit area of cross-section) we permit, as distinct from the stress which would break the material, is called the **working stress** of the material, and the ratio $\frac{\text{Ultimate stress}}{\text{Working stress}}$ is called the **factor of safety.**

The factor of safety depends on many things: cost, uncertainty of conditions, consequences of a possible failure, etc. For example, in structural steelwork, the factor is usually about 3. In the design of steam boilers it is much higher, perhaps between 8 and 10. The probable reason is that the possibility of an accidental high overload is much higher than in a steel building, and the consequences of a failure might be disastrous. In an aircraft, where it is of great importance to keep weights down as much as possible, the factor of safety may be less than 2 in some applications.

(3) Uniform and Non-uniform Stress

In the reasoning of Section 1 of this chapter we made a very important assumption upon which the whole notion of stress is based: namely that the material withstands load in proportion to its cross-sectional area. This proposition is not quite so justifiable as it may seem, because it rests on the further assumption that every point of the cross-section behaves exactly as every other point, or in other words, that the stress is distributed uniformly over the cross-section. But this assumption is by no means always valid.

Fig. 78 shows three examples of load applied to a bar where it would be wrong to assume a uniform distribution of load over the whole cross-section. Fig. 78 (*a*) shows load applied to a bar by means of pins and shackles at each end. Imagine the bar to have a cross-sectional area of 1 square centimetre, and to be made of steel. Then we might assume that it would break when the load reached a value of 45 kilonewtons. But it does not require much specialized judgment to see that it will actually break at a lower load than this. Although we may fairly assume the load to be evenly shared out over the cross-section at *A–A*, we certainly are not justified in assuming such a state at the section *B–B* near to the pins. The corners, which I have shown shaded, manifestly take no load at all, so the stress will be correspondingly higher over the remaining part of the section. Most probably, the bar would break at some load less than 45 kilonewtons, with the pin anchorage tearing out of the end.

Fig. 78 (*b*) shows a less obvious type of non-uniform load. There are no holes this time to break the uniformity of the cross-section, but the load is applied to the bar by gripping it in jaws at the two ends. In such a case, the bar is liable to break near one or other of the jaws. Here, the non-uniformity is due to the sideways grip of the jaws on the material. For this reason, bars submitted to a tensile test in a machine are specially shaped for the purpose, with the ends enlarged as shown.

Fig. 78 (*c*) is similar. One might feel justified in taking 1 square centimetre as the minimum area of the non-uniform bar, but in practice you would again find that the bar broke close to the junction with the larger section, and at an average value of stress less than the ultimate. The sudden change of section causes a non-uniformity of load over the cross-section close to the junction.

Holes in bars, and sudden changes of section are known to stress-men as 'stress-raisers', and are to be avoided whenever possible. A sudden change of section can usually be replaced by a gradual change, as shown in the lower part of Fig. 78 (*c*).

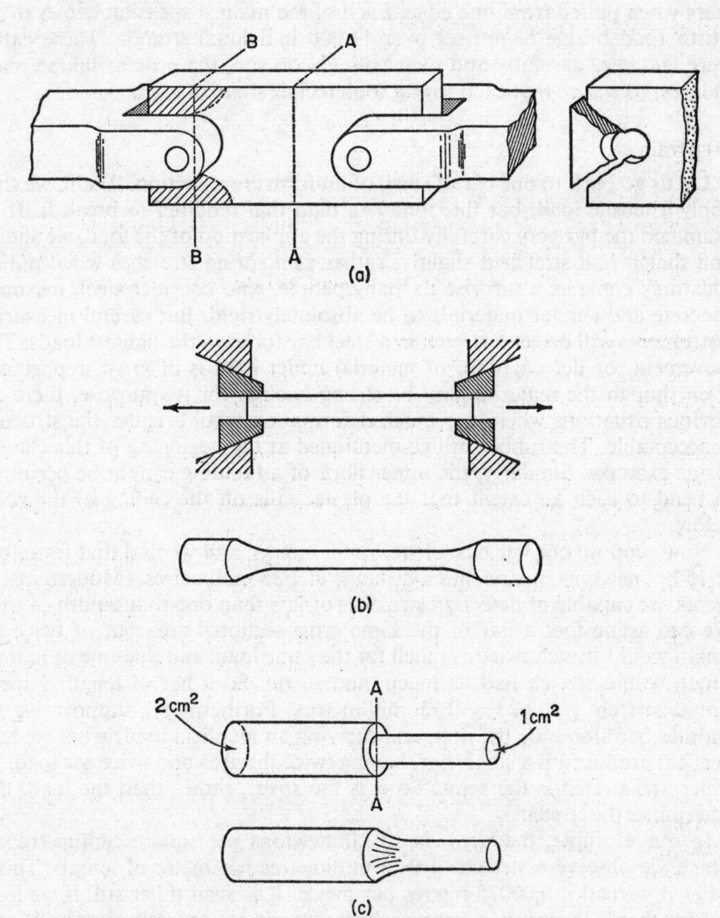

Fig. 78. Three examples of non-uniform stress distribution.

As a last example of possible non-uniformity, consider a steel cable or hawser made up of many individual wires stranded together. Depending on how the cable is made, and used, there is always the possibility that various strands might be under-loaded, with the result that others may become over-loaded. The situation can be likened to 20 boys pulling a rope in a tug-of-war. Suppose five boys decide to take it easy for a few seconds, and leave it to the other 15. These 15 are then overcome, and collapse. The five now decide to take their share, but it is too late: the other 15 have 'broken', so to speak.

The five are left with the full load, which soon causes them in turn to be overcome. This situation could not occur if all 20 boys pulled steadily and continuously, as every tug-of-war captain must know. So, a badly-laid stranded cable might rupture under load, very much as a piece of perforated paper tears when pulled from one edge. Each of the main suspension cables of the Forth road bridge comprises over 11 000 individual strands. These cables were laid very carefully and painstakingly on site, the process taking many months, to ensure that each strand took its fair share of load.

(4) Strain

Let us go back to our bar of steel, of uniform cross-section. Again, we shall apply a tensile load, but this time less than that required to break it. If we examined the bar very carefully during the application of the load, we should find that it had stretched slightly, rather as a spring stretches when pulled. This may come as a surprise to many people, who consider steel, masonry, concrete and similar materials to be absolutely rigid. But careful measuring instruments will detect a stretch in a steel bar for even the lightest loads. This movement, or deformation, of material under load is of great importance. Even though the material may be strong enough for its purpose, there are obvious situations where too much deformation would render the structure unacceptable. The rubber bridge mentioned at the beginning of this chapter is one example. Similarly, the upper floor of a building cannot be permitted to bend to such an extent that the plaster falls off the ceiling of the room below.

Now suppose our bar has a length of 3 metres, and we find that for a load of 15 kilonewtons, it stretches a distance of 2·25 millimetres. (Modern instruments are capable of detecting stretches of less than one-thousandth of this.) We can argue that a bar of the same cross-sectional area but of twice the length would stretch twice as much for the same load, and that one of half the length would stretch half as much, and so on. So a bar of length 1 metre would stretch $\frac{1}{3} \times 2{\cdot}25 = 0{\cdot}75$ millimetres. Furthermore, suppose we lay another bar alongside the first, and carrying an identical load; what we have in effect produced is a single bar, having twice the area and twice the load, for which the stretch is the same. So it is the *stress*, rather than the load, that determines the stretch.

In our example, the stress is 15 kilonewtons per square centimetre, for which we observe a stretch of 0·75 millimetres per metre of length. This is tidier if we call it 0·00075 metres per metre. It is even tidier still if we leave out the words 'metres per metre', which you can see are unnecessary. If each metre stretches 0·00075 metres, then each centimetre will stretch 0·00075 centimetres, each inch will stretch 0·00075 inches, and so on. The figure 0·00075 is a proportional, or fractional, stretch for this value of stress; it need not be expressed in units. This proportional stretch is called the **strain.**

Although I have referred constantly to a tensile (i.e. pulling) load, the same concepts of stress and strain are valid for compressive (i.e. pushing) loads.

(5) The Stress–Strain Graph

So far we have applied a breaking load to a material, and we have applied a moderate load and measured the stretch. This has enabled us to do two things. First, we can decide a maximum load for any bar of our material, based on a fraction of the 'ultimate stress'. Secondly, we can predict the stretch of any bar under a load, *provided* the stress agrees with the value of stress in our test. In numbers, if we stress a steel bar to a value of 15 kilonewtons per square centimetre, we know that the strain will be 0·00075, and we can work out the stretch from this. But what if the stress is different? Clearly, what we have to do is to extend the range of our experiment and apply not one load but a range of loads.

Suppose we take a bar of a certain material and perform a careful test upon it, observing the stretch over a measured length for various carefully measured loads. This is known as a tensile test. The results of a tensile test are given in the first two columns of the following table.

RESULTS OF TENSILE TEST

Cross-sectional area of bar = 4 cm²; gauge length = 10 cm

Load	*Stretch*	*Stress* $= \dfrac{Load}{Area}$	*Strain* $= \dfrac{Stretch}{Gauge\ length}$
kN	mm	kN cm^{-2}	
0	0	0	0
5	0·005	1·25	0·00005
10	0·01	2·5	0·0001
15	0·016	3·75	0·00016
20	0·0225	5·0	0·000225
25	0·0295	6·25	0·000295
30	0·0375	7·5	0·000375
35	0·047	8·75	0·00047
40	0·060	10·0	0·0006
45	0·078	11·25	0·00078
50	0·105	12·5	0·00105

The third column of the table gives the stress, i.e. the load divided by the area of cross-section; in this example the area of cross-section is 4 square centimetres. The fourth column shows the strain. The 'gauge length' stated at the top of the table is the length over which the stretch is actually measured. This measurement is usually made by clamping an instrument called an extensometer (Fig. 80) on to the bar. The extensometer is designed to measure the stretch (or compression) between two points on the bar which are a distance apart equal to the gauge length. Thus the strain for each reading is the stretch divided by the gauge length.

We can now produce a graph of stress against strain, as shown in Fig. 79. The important point to appreciate at this stage is that the graph represents a property of the *material* of which the bar is made, rather than a property of the bar itself. It can be used to determine the stretch of any other bar of the same material, for any load, always providing the stress produced is within the stress range of our test. Let us determine the stretch of a piece of wire

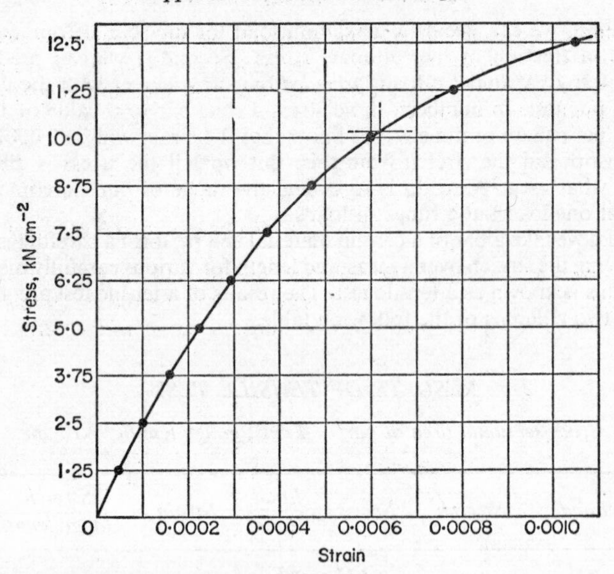

Fig. 79. Stress–strain graph for an unnamed material.

made of this material, 0·5 millimetres in diameter, of length 8 metres, and carrying a load of 20 newtons.

First we evaluate the stress. Designating this by f:

$$f = \frac{\text{load}}{\text{area}} = \frac{20}{\pi/4 \times (0{\cdot}5)^2}$$

$$= 102 \text{ newtons per square millimetre}$$

$$= \frac{102}{1000} \times 10^2 \text{ kilonewtons per square centimetre}$$

$$= 10{\cdot}2 \text{ kN cm}^{-2}$$

Referring now to our graph (Fig. 79), we find that the strain corresponding to this stress of 10·2 kN cm^{-2} has a value of 0·00062. Since

$$\text{strain} = \frac{\text{stretch}}{\text{length}},$$

we have

$$\text{stretch} = \text{strain} \times \text{length}$$

$$= 0{\cdot}00062 \times 8 \text{ metres}$$

$$= 0{\cdot}00496 \text{ metres}$$

$$= 4{\cdot}96 \text{ millimetres}$$

Observe that although the strain has no units, the stretch is calculated in the units of length chosen (metres in this case).

To illustrate the limitation of such a graph, we shall now try to evaluate the stretch of a bar of the same material having a circular cross-section of 2·5 centimetres diameter. The bar is 10 metres in length, and is subjected to a tensile load of 100 kilonewtons.

$$f = \frac{\text{load}}{\text{area}} = \frac{100}{\pi/4 \times (2{\cdot}5)^2}$$

$$= 20{\cdot}4 \text{ kilonewtons per square centimetre}$$

If we now refer to Fig. 79, we see that this stress is too high, being outside the range of the graph. So for this case we cannot use the graph to determine the stretch. Indeed, the load might be sufficient to break the bar.

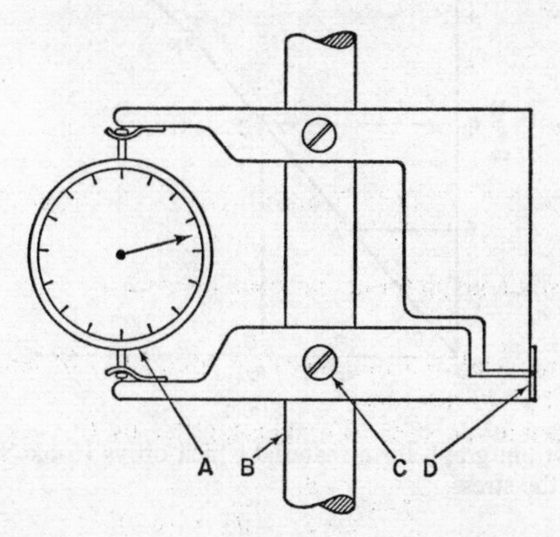

Fig. 80. A simple type of extensometer. The instrument is clamped on to the specimen under test (*B*) by two screws (*C*) which are a set distance apart equal to the gauge length. The framework is in two parts, hinged together at (*D*). Elongation of the specimen causes the jaws at the left-hand end to open a small amount. This movement is measured by the dial gauge (*A*), which is a simple gear mechanism with a pointer giving a high magnification of movement.

(6) Hooke's Law

The stress–strain graph of Fig. 79 is drawn for an imaginary material, merely introduced to show how such a graph can be used. If we were to perform an actual experimental tensile test on some modern materials, such as one of the plastics, we might well obtain a very similar type of graph. But a test on steel, or on an aluminium alloy, or even on wood, would give us a stress–strain graph which would be almost a perfect straight line, for the initial stages of loading. The strain would be proportional to the stress. Fig. 81 shows such a graph.

This linear stress–strain relationship was discovered by Robert Hooke, in 1678. When a material exhibits a stress–strain graph which is a straight line like this, it is said to obey Hooke's Law. All materials fail to conform to this law after a certain limiting stress is reached, and the graph becomes curved.

The point at which this occurs, shown on Fig. 81 as P, is called the **limit of proportionality**.

If a material conforms to Hooke's Law, the mathematical analysis of many examples of stress is often relatively simple. The analysis of stress due to bending is a notable example, which we shall examine in the next chapter. For materials which do not conform to this law, other, and often more elaborate, methods of analysis may have to be used.

If we wish to predict the deformation of a bar due to a tensile load, and we know from experience that the material obeys Hooke's Law, we do not now need even to draw the graph. All we need is the slope of the graph, or the

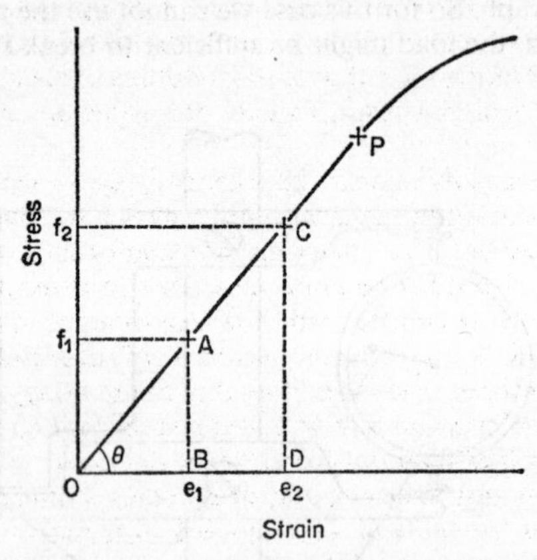

Fig. 81. Stress–strain graph for a material which obeys Hooke's Law: the strain is proportional to the stress.

ratio of stress/strain. You can see from Fig. 81 that, for a straight-line graph this ratio must be the same for any value of stress. I have shown two values of stress, f_1 and f_2, and the corresponding values of strain, e_1 and e_2. The two ratios of stress/strain are respectively f_1/e_1 and f_2/e_2: in geometrical terms, AB/OB and CD/OD. These ratios are the same, because the triangles OAB and OCD are similar. For the trigonometrically minded reader, the ratio is the tangent of the angle θ.

(7) Modulus of Elasticity

The constant ratio of stress/strain for a material which obeys Hooke's Law is called the **modulus of elasticity**, or **Young's modulus**, and is usually designated by E. Denoting stress and strain by f and e, we can state:

$$E = \frac{f}{e}$$

But stress can be expressed as load divided by area, or, algebraically as $f = F/a$. Similarly, strain can be expressed as the stretch, or elongation, divided by the length, or, algebraically as $e = x/L$. Then:

$$E = \frac{f}{e} = \frac{F/a}{x/L} = \frac{FL}{ax}$$

This simple formula enables us to calculate the elongation of a bar under any load for a material obeying Hooke's Law, providing the load is within the limit of proportionality for the material. For steel, E has an approximate value of 200×10^9 newtons per square metre, or 200 giganewtons per square metre, usually written in the abbreviated form $200\ \mathrm{GN\ m^{-2}}$. For most alloys of aluminium, the value is about one-third of this. Stress-analysis textbooks and designers' handbooks give values for all commonly-used materials.

I should point out that, although all the previous discussion is based on the behaviour of material under a tensile (pulling) load, exactly the same behaviour is observed in material subjected to a compressive (pushing) load, with the same reservations concerning a uniform load distribution mentioned in Section 3. Furthermore, the value of the modulus of elasticity will be found to be, for all practical purposes, exactly the same in compression as in tension.

The modulus of elasticity appears in practically every aspect of the loading and resultant deformation of materials, and makes a relatively simple mathematical analysis possible. For materials not conforming to the law, much more experiment and testing becomes necessary to determine deformations of members and structures under load. E is best thought of as an index of the **stiffness** of a material. A high E-value means a stiff material. Thus, steel may be considered about three times as stiff as aluminium alloy—a fact which may help to explain why aluminium alloys have never been used much in building. Steel is one of the stiffest materials used by the engineer.

Being the ratio of stress to strain, and strain being a dimensionless fraction, it follows that E has the units of a stress itself. It is the stress to cause theoretical unit strain in the material, hence it may be defined as the theoretical stress which would cause a bar of the material to extend to double its length. We have to include the word 'theoretical' because, for most materials, the bar would break long before such a state was reached. A possible exception is soft rubber, which could quite conceivably double its length before breaking.

As a simple example of the use of E, let us consider a railway signal connected to the control box 500 metres away by a steel wire of cross-sectional area 20 square millimetres. Assume that the force required to actuate the signal is 300 newtons, and that the end attached to the signal is required to move 9 centimetres. How far shall we have to pull the operating end?

The elongation of the wire is given by our formula. Rearranging and substituting values:

$$x = \frac{FL}{aE}$$

$$= \frac{300 \times 500}{(20 \times 10^{-6}) \times (200 \times 10^{9})}$$

$$= 0{\cdot}0375 \text{ metres} = 3{\cdot}75 \text{ centimetres}$$

(Notice how all terms must be reduced to consistent units of newtons and metres.)

So, in order to move the signal end 9 centimetres, we should have to pull the other end of the wire $(9 + 3{\cdot}75) = 12{\cdot}75$ centimetres.

(8) Stress in a Thin Cylinder

A thin cylindrical shell subjected to an internal pressure affords an illustration of a fairly simple stress calculation. An actual example of such a cylinder would be a steam-locomotive boiler. Fig. 82 shows the cylinder, with an imaginary cut made in it so that we can look inside. It would be the job of a designer to specify the thickness of the shell wall, given the diameter of the cylinder and the internal pressure.

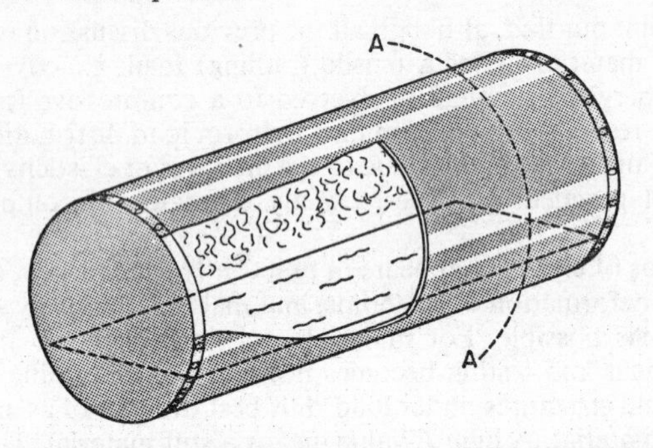

Fig. 82. A steam boiler. The cylindrical wall is stressed by the pressure inside.

The method of attack with practically all stress problems is to make an imaginary section through the loaded member, and then to treat one of the sectioned halves as a free-body diagram. Let us perform this trick on the cylinder. We will make an imaginary cut along the line *A–A* and then examine the equilibrium of forces on the right-hand end which is isolated by the cut. The end by itself is shown in Fig. 83.

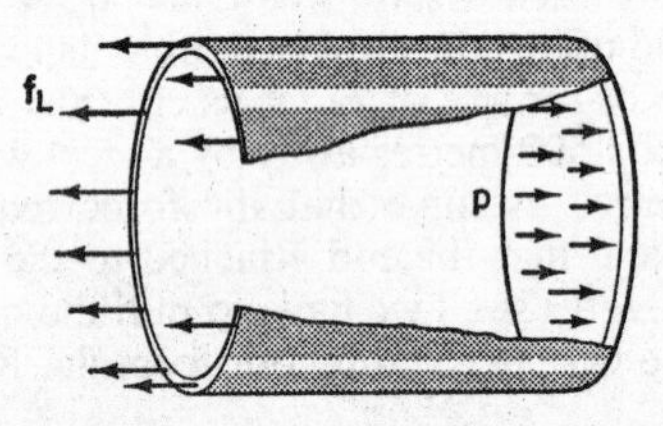

Fig. 83. A circumferential 'section' through the cylindrical shell. The pressure force acting to the right is balanced by the stress force acting to the left.

What forces are acting on this 'cut' section of boiler shell? The internal pressure acts on the end-plate, with a force tending to drive it from left to right. But the portion of shell is in equilibrium, because it is not moving to the right in obedience to this force. The pressure force must therefore be balanced by an equal force in the opposite direction. This force is supplied by the tensile stress acting round the plane of section. In physical terms, the pressure force on the boiler end is balanced by the equal and opposite force exerted *on* the end *by* the remaining portion of the boiler: the end is, so to speak, being pulled to the left by the other half.

We can calculate the stress over this section by equating these two forces. If we call the stress f_L, the shell diameter D and the plate thickness t, the force due to the stress is (stress × area). It is sufficiently accurate to assume that the stressed area is $t \times \pi D$ so that the stress-force is $f_L \times t \times \pi D$. The pressure force to the right is (pressure × area) $= p \times \pi/4D^2$. (The fact that there may also be water in the boiler will not materially affect this force.) The two forces are equal and opposite:

$$f_L \times t \times \pi D = p \times \frac{\pi}{4} D^2$$

which gives:

$$f_L = \frac{pD}{4t}$$

If the pressure increases to such a value that f_L reaches the ultimate stress of the material, we may expect the boiler to burst by splitting along a circumferential line, the two halves being blown apart from each other. This has never happened in recorded history—not because of the great care taken in the design and use of boilers, but because failure is far more likely to occur in another manner, as we are about to show.

Consider now the possibility of the upper half of the boiler parting from the lower half. To investigate this possibility, we make an imaginary section, not along a circumferential line as before, but along a lateral line, as in Fig. 84. Taking the lower half now as our free body, the pressure acts radially all over the inside of the shell, giving a resultant force downwards. This resultant force is the product of the pressure and the **diametral** area, i.e. the projected area of the curved surface. You may appreciate this better if you imagine the lower half of the shell to be filled with concrete up to the half-way line. The

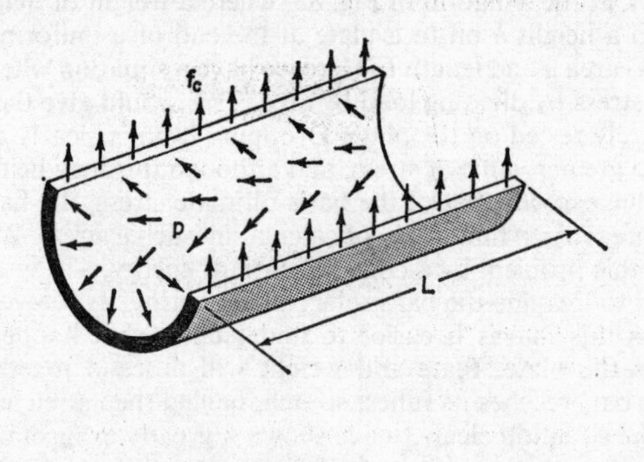

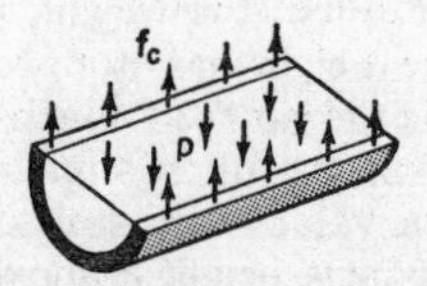

Fig. 84. A lateral section through the boiler shell. The pressure-force acting downwards is balanced by the stress-force acting upwards

pressure would then act vertically downwards over the whole of the flat upper concrete surface, and the force would be $p \times DL$ where L is the length of the shell. Common sense urges that the introduction of concrete into a pressurized cylinder is not going to affect the stress in the wall of the cylinder, if the pressure remains the same.

The stress-force, or the force with which the upper half pulls the lower half up, is the product of the circumferential stress f_C and the double wall-area $2 \times tL$. Equating the two forces:

$$f_C \times 2tL = p \times DL$$

which gives:

$$f_C = \frac{pD}{2t}$$

The result is similar to the previous one, except that the value of the stress is twice as much. So the boiler wall is actually experiencing two stresses at the same time in two directions at right-angles: a longitudinal stress of magnitude $pD/4t$, and a circumferential stress of twice this amount. Therefore, if it is going to burst, it will do so not along the circumferential line of least stress, but along the longitudinal line of greatest stress.

The formula also tells us that if we want to reduce the stress for a given pressure, we can either increase the thickness (which is obvious) or decrease the diameter (which is not quite so obvious). Modern high-pressure boilers are never more than about 2 feet in diameter; the eight-foot giants of Victorian times were mostly designed for very much lower pressures.

(9) Impact Loads

A special and important case of stress arises when a load is applied suddenly to a bar. Look at the situation in Fig. 85, where a weight of magnitude W is dropped from a height h on to a plate at the end of a uniform steel bar of cross-sectional area a and length L. Here we have a situation where we cannot calculate the stress by dividing load by area. This would give the stress if the weight W merely rested on the plate. Dropping it on is clearly going to give rise to a much greater value of stress, and although this may be only momentary, if its value exceeds that of the bar's ultimate stress, the bar will break. It becomes necessary to find a way of calculating such a stress. We can obtain an answer to this problem by a consideration of energy.

It may help to imagine the bar replaced by a spring, as I have indicated in Fig. 85 (*b*), as this makes it easier to understand what happens when the weight strikes the plate. Plate and weight will descend together until the spring (or the bar) reaches its fullest stretch, having then stretched a distance y. To make the situation clear, I have shown a greatly exaggerated stretch y. At this point, two things are clear. First, the weight, plate and spring are instantaneously at rest, because the weight has stopped going down and is about to come up. Secondly, this is the point at which the stress in the bar has its greatest value, because it is at the fullest stretch. This is the instant, then, when we need to determine the value of the stress.

Before the weight is released it is at a height h above the plate. When it comes to rest it is at a depth y below the original position of the plate. So it loses potential energy of an amount $W(h+y)$. At the lowest point, everything is at rest, so there is no kinetic energy to account for this loss of potential energy. How, then, do we account for it? The answer is that the energy has been expended in stretching the bar or the spring. The material is, so to speak, in a condition of suppressed or latent energy, capable of being released.

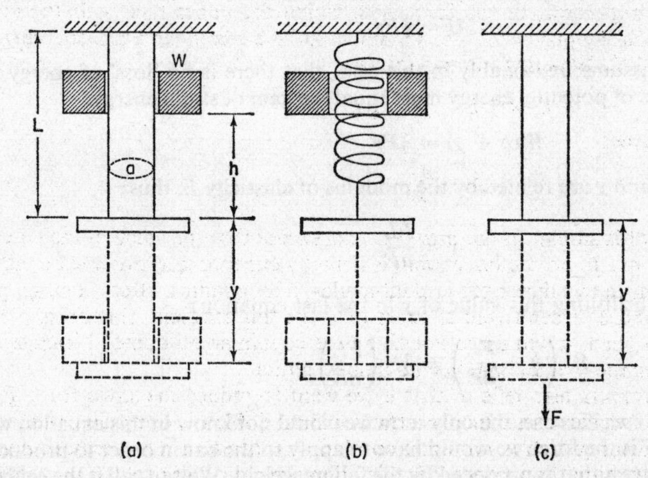

Fig. 85. (*a*) A weight is dropped on to a bar from a height h; the bar stretches a distance y. (*b*) The bar may be regarded as a very stiff spring. (*c*) The same stress and extension would be produced by a force F applied steadily.

It is rather like potential energy, except that the bar or spring itself remains where it is and is not raised up. This type of energy absorbed in deforming material is called **strain energy**. It is easy to work out its value in our example.

Imagine stretching the bar the same amount, but by means of a steadily increasing force: the energy absorbed by the bar will be the work done by this force. The bar will stretch as the force increases. If the bar is made of a material which obeys Hooke's Law, a graph of the force against the stretch will be a straight line. Let us draw this graph, as shown in Fig. 86.

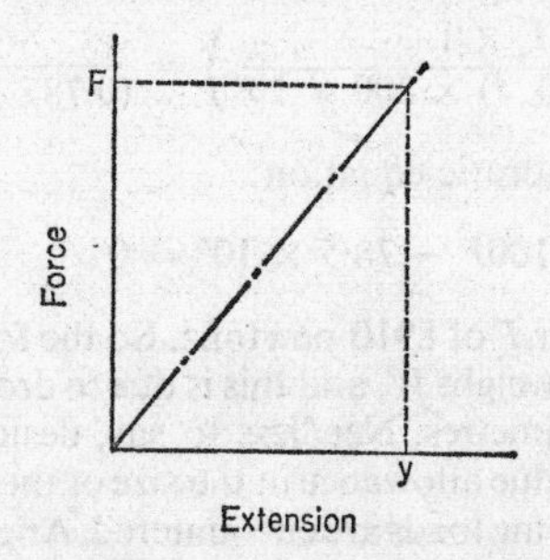

Fig. 86. The force–extension graph for a uniform bar being stretched. The work done U is the product of the average force $\frac{1}{2}F$ and the stretch y. $U = \frac{1}{2}Fy$.

Assume that when the bar has stretched an amount y, the corresponding force is F. The work done by the stretching force is not $F \times y$, because F has not been applied for the full distance. The work done is the *average* force times the distance, and the average force in this case is clearly $\frac{1}{2}F$. Calling the work U:

$$U = \tfrac{1}{2}Fy$$

If we assume, reasonably in this case, that there is no 'loss' of energy, then the loss of potential energy must equal the gain of strain energy.

$$W(h + y) = \tfrac{1}{2}Fy$$

But F and y are related by the modulus of elasticity E, thus:

$$y = \frac{FL}{aE}$$

and substituting this value of y in the last equation:

$$W\left(h + \frac{FL}{aE}\right) = \tfrac{1}{2}F\left(\frac{FL}{aE}\right)$$

In any specific case, the only term we would not know in this equation would be F. F is the force we would have to apply to the bar in order to produce the same stress that is produced by the falling weight. We can call it the **equivalent static load.**

The importance of this problem is best illustrated by an actual example. Suppose the bar to have a diameter of 1 centimetre and a length of 1 metre. Let W be 50 newtons. This is the approximate weight of a mass of 5 kilogrammes—a very modest load for such a bar. We will drop the weight from a height of 5 centimetres, and for E we will take our previous value of 200×10^9 newtons per square metre. The cross-sectional area of the bar is $\pi/4 \times 1^2 = 0{\cdot}785$ squre centimetres.

$$W\left(h + \frac{FL}{aE}\right) = \frac{1}{2}\frac{F^2L}{aE}$$

Substituting all the given values:

$$50\left(0{\cdot}05 + \frac{F \times 1}{(0{\cdot}785 \times 10^{-4}) \times 200 \times 10^9}\right) = \frac{\frac{1}{2}F^2 \times 1}{(0{\cdot}785 \times 10^{-4}) \times 200 \times 10^9}$$

This reduces to the quadratic equation:

$$F^2 - 100F - 78{\cdot}5 \times 10^6 = 0$$

This gives a value for F of 8910 newtons. So the force F is some 178 times the value of the original weight W, and this is due to dropping the weight merely from a height of 5 centimetres. Needless to say, designers faced with such a possibility would make due allowance in the size of the bar. Similar conditions are found whenever moving loads are encountered. Another instance is that of a vehicle or a ship being towed. If the designer merely specifies the towline on the basis of the steady pull required, he is asking for trouble. Every time the slack line is suddenly tightened there will be a stress possibly many hundreds of times greater than that of the steady pull. Oddly enough, this stress can be reduced by making the rope *longer*. It will then stretch further for a given load, and thus will accept more energy than a shorter rope.

(10) Shear Stress and Strain

Tension and compression represent only two ways in which material may be loaded. It may be subjected to shear, to bending, to twisting, to pressure, to buckling, or to various combinations of these. We shall deal with bending and twisting in the following two chapters, and we shall see that bending consists of a combination of tensile and compressive load. Twisting, or torsion, is an example of shear stress and strain, and we must examine what this type of stress does to a material.

Shear stress is a transverse stress *across* a plane of material, as distinct from a direct, or normal, stress perpendicular to the plane. (Normal, in this context, means perpendicular.) Its effect on the material is best visualized by laying a thick book flat on the table, placing your hand on top of it, and pushing horizontally, as in Fig. 87. The effect is to make adjacent pages slide

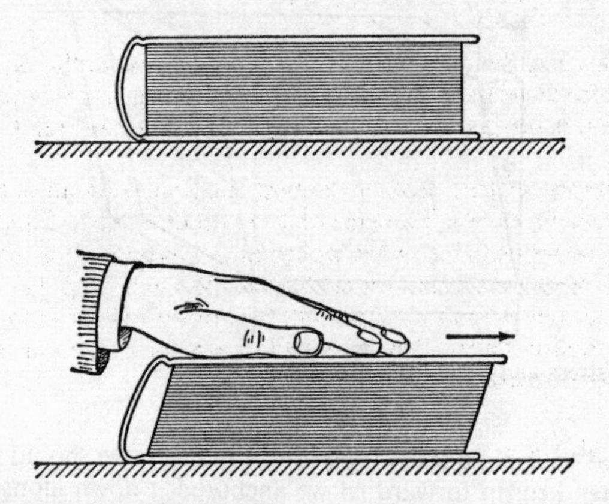

Fig. 87. A representation of shear strain.

over one another, so that the whole book becomes a parallelogram in end view instead of a rectangle. A similar effect, although much less obvious, attends a transverse shear load in a material.

Fig. 88 shows a diagrammatic representation of material subjected to a shear load, with the resultant strain greatly exaggerated. The original configuration of the block is shown dotted. The lengths of the sides of the block are a, b and c. We first define the stress, or load intensity, q as load/area:

$$q = \frac{F}{ab}$$

The shear strain is not quite so easy to define. The actual deformation of the block is x, and the 'length' in this case can be thought of as the height c. So the deformation per unit length corresponding to the tensile or compressive strain is x/c. For all engineering materials in common use, this ratio is always extremely small, and is practically equal to the angle φ (Greek phi) measured in radians.

A careful experiment under ideal conditions would give a straight-line graph of shear stress against shear strain for many engineering materials, exactly as for tensile stress and strain. That is to say, Hooke's Law is valid for many materials in shear as well as in tension. The ratio q/φ would be constant up to a certain limit of proportionality. This constant ratio is another modulus of the material; it is called the **modulus of rigidity**, and usually denoted by G. Thus:

$$G = \frac{q}{\varphi}$$

For steel, G has a value of approximately 80×10^9 newtons per square metre.

Although this relationship is most conveniently demonstrated with the aid of Fig. 88, I must emphasize that it would be impossible to observe in prac-

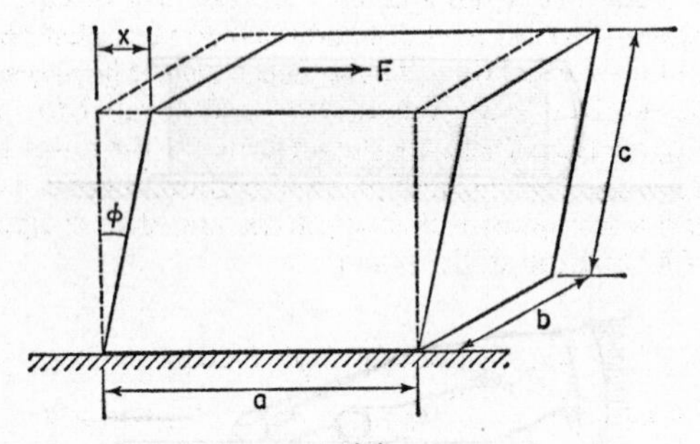

Fig. 88. Shear stress and strain in a material.

tice. If we tried to apply a shear load in this manner, we should find that the block would tend to tip forward; if we anchored it down along the bottom face, this would give rise to tensile and compressive stresses. As I have said, the only practical way of achieving pure shear is by twisting a circular bar, but the geometry is then not so easy to illustrate.

(11) Complex Stress

You may consider a simple tensile load as devoid of any complication, and merely giving a simple tensile stress which is the intensity of load over the cross-section of the loaded member. If we take a bar of uniform cross-section, make a 'section' through it, and examine the equilibrium of one half, we can equate the stress-force over the 'cut' face to the external load, as shown in Fig. 89.

The equation of equilibrium is:

$$F = f_T \times a$$

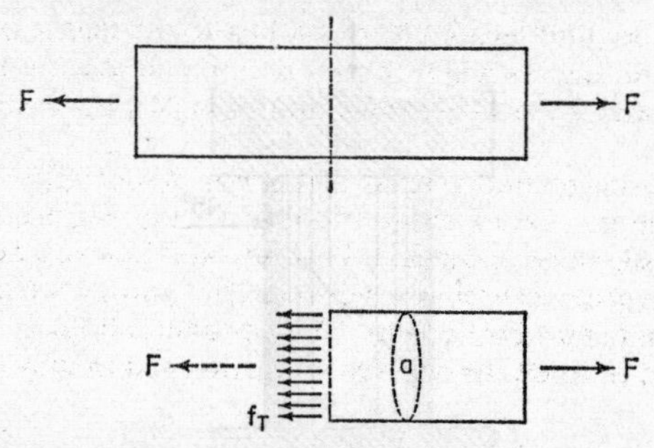

Fig. 89. Simple tensile stress over a normal cross-section.

Therefore

$$f_T = \frac{F}{a}$$

This is simple. But suppose that instead of making a section perpendicular to the axis, we made one at an angle, as in Fig. 90. For equilibrium of the 'cut' half, the stress-force on the left-hand inclined face must be exactly equal and opposite to the external force *F*, and hence *cannot be perpendicular to the face*. It may be resolved into components, one parallel to, and the other perpendicular to, the face. These are shown dotted in the lower half of Fig. 90. In other words, on such an inclined plane of section, there is a component of tensile stress and a component of shear stress, and their magnitudes depend on the angle of inclination of the plane.

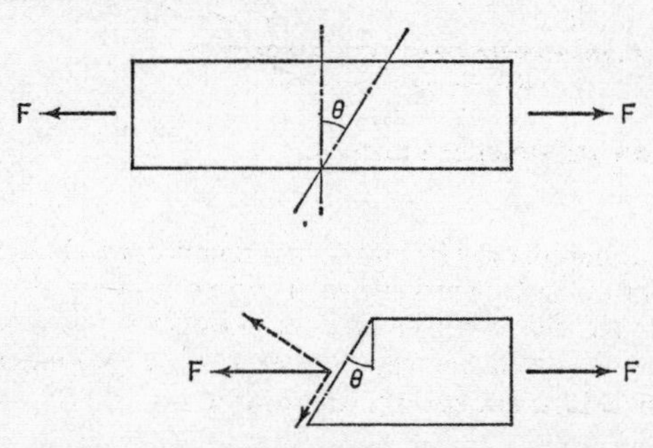

Fig. 90. A direct tensile load causes both direct and shear stress on an oblique plane.

This is merely an introduction to a large section of stress analysis called **complex stress**, and it is not possible here to take it any further. The essential fact to realize is that, for a material subjected to any load, the stresses comprise both direct (i.e. tensile or compressive) and shear, and the relative magnitudes of the stresses depend on the plane you are examining. That this is not mere mathematical trickery is shown by the way some materials fracture. If you take a brittle substance like cast iron, brass or concrete, and crush it (i.e. apply a direct compressive load to the ends), the material will invariably break along a line approximately 45° to the axis of load. Fig. 91

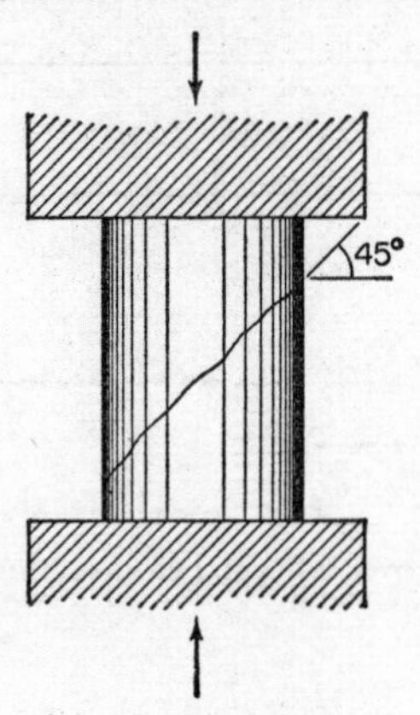

Fig. 91. Compression failure of cast iron.

illustrates the failure of a short cylinder of cast iron. It can be shown that this is the plane along which the shear component of stress has its maximum value. The type of failure is evidence that the material has attained a critical value of *shear* stress along this plane, although the applied load is a direct load and not a shear load.

CHAPTER NINE

STRESSES DUE TO BENDING

We must examine this method of subjecting material to a load for two reasons. In the first place, bending forms one of the most commonly encountered forms of load-carrying: every floor of every building is subject to bending; likewise, every crane, and every bridge. Secondly, of all the possible ways material may be called upon to carry a load, bending is probably the most destructive. By this, I mean that failure of any given component is most likely to occur if the load is applied in the form of a bending load. If asked to break a walking-stick, you would presumably hold it by the ends and break it (or try to) across your knee. You would be unlikely to hook one end around a post and apply a direct pull to the stick.

A bar of steel having a square cross-section of area 1 square centimetre is capable of withstanding a load in direct tension of approximately 30 kilonewtons (or 7000 pounds) before the stress may be said to have reached a dangerously high value. But if one end of such a bar were placed in a vice, and a transverse load applied at, say 1 metre from the fixed end, a load of perhaps 50 newtons (or 12 pounds) would be sufficient to produce the same value of stress. Furthermore, if we took such a bar, and rested it on two supports about 7 metres apart, like a bridge, we should find that the weight of the bar alone would be sufficient to produce the same value of limiting stress as in the first two cases. Although it is quite common to see bridges which are designed according to this principle, with the load being carried by horizontal girders supported only at the two ends, you may judge by looking at such a bridge that a good proportion of the load it has to support is due to its own weight. For this reason, such a design cannot be used for anything other than very moderate spans of say 20 to 30 metres. For greater spans, either intermediate supports have to be provided, or a design has to be used which gives rise to a much smaller bending effect.

Fig. 92 shows a type of bridge used for large spans. The actual load-bearing

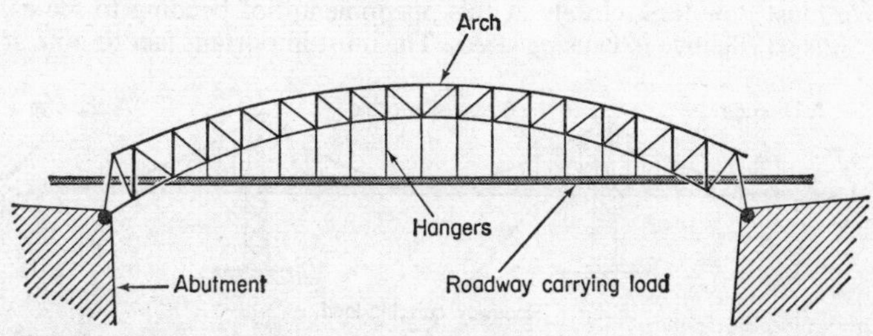

Fig. 92. An arch bridge. The roadway is hung from the arch, thus avoiding large bending moment on the structure.

roadway is horizontal and relatively light in construction, and is hung by vertical rods from the main supporting structure, which is a steel arch. Because of the shape of the arch, the transverse load of the roadway, which would give rise to bending of a horizontal girder, causes instead a compression thrust along the length of the arch. Heavy abutments must be provided to withstand this thrust.

The same principle applies to a brick arch, such as an old railway bridge. The roadway or trackway carrying the load is in this case entirely above the crown of the arch instead of being slung below it. You can see in Fig. 93 how the downward thrust of the load is transformed into a compressive thrust

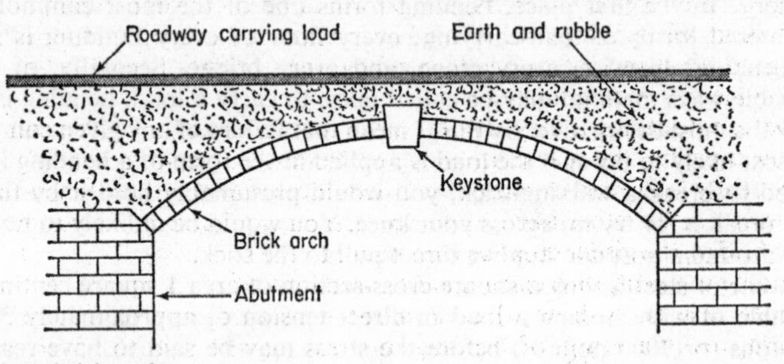

Fig. 93. A brick arch.

between the bricks of the arch. At the crown of the arch is the keystone, beloved of imaginative writers as representing the whole strength of the arch. The keystone actually takes no more load than any other brick in the arch. If its removal would cause the possible collapse of the whole structure, the same is true of any other part of the arch.

Fig. 94 shows the principle of the arch turned upside-down, so to speak. The transverse load is hung from what amounts to an inverted arch, giving rise to a tensile load instead of compression. The inverted arch comprises a number of steel cables or chains. Most readers will recognize the familiar outline of the suspension bridge. A modest example is the Clifton suspension bridge in Bristol, spanning 702 feet. The new bridge spanning the Forth estuary for a distance of 5980 feet, is a more spectacular example.

(1) Bending Moment

We must now look closely at this phenomenon of bending to see exactly why it is so effective in causing stress. The most important fact to note at this

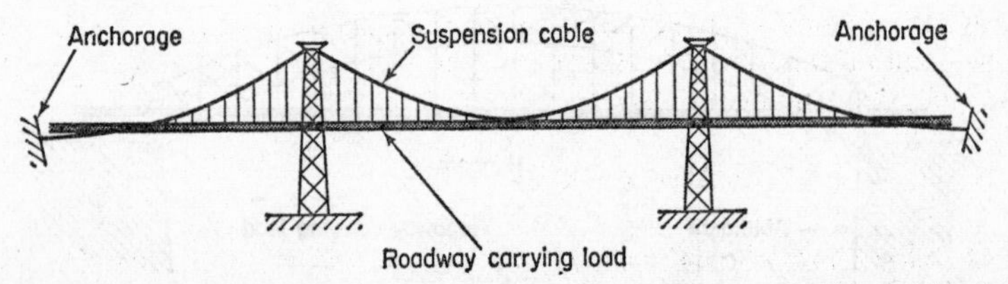

Fig. 94. A suspension bridge. The roadway is hung from the cable, which is in tension.

stage is that it is not merely the value of the load which determines the stress, but also the position where the load is applied. In breaking a walking-stick across your knee, you would presumably grasp the stick as near to the ends as possible, not close to the knee. It is very much akin to the turning effect of a force we examined in Chapter Three, except that in this case the turning effect is resisted by the strength of the material. This turning effect of the load is called **bending moment,** and it is evaluated, as was our earlier moment, by multiplying force by distance. If a number of forces combine on a beam, the bending moment at any point is determined by adding together the individual moments due to each load, not forgetting that some may be in the opposite direction to others, and so would be negative. Bending moment is to a beam what direct load is to a bar in tension. In order to determine the maximum stress in a beam due to bending, the first task is to find the maximum value of bending moment.

If we are designing a very simple type of beam, having a fairly small load-bearing capacity, the maximum value of bending moment is the only value we are really interested in, since it will determine the minimum dimensions of the beam. On the other hand, if we are designing a large structure such as a girder bridge, the girder cross-section may be varied to allow for the variation in bending moment, thus using material economically. The regular procedure in either case is to draw what is called a **bending-moment diagram.** This diagram, really a graph of bending moment against distance from end of beam, is always drawn below a diagram of the loaded beam so that the value of bending moment at any point can be immediately seen. Unless the loading of the beam is extremely simple, the drawing of the bending-moment diagram is sound practice, even if only a maximum value is required; determining the location and value of maximum bending moment by mere inspection can lead to mistakes. Before proceeding to determine the relationship between bending moment and bending stress, we shall examine a few simple cases of bending moment.

First, let us take the case of a single load W on a beam which is supported at its two ends. The beam diagram is shown in Fig. 95, together with the bending-moment diagram. It is seen that the maximum bending moment has the value Wab/L. This expression is easily verified. We first evaluate the left-hand reaction R_1 by taking moments of forces about the right-hand end. The equation is:

$$R_1 \times L = W \times b$$

$$R_1 = \frac{Wb}{L}$$

The bending moment at the point where the load is applied is the sum of moments of all forces to one side of the point chosen. If we choose the left-hand side, the bending moment is:

$$M = R_1 \times a = \frac{Wab}{L}$$

The decision as to whether this is positive or negative is arbitrary. On the right of Fig. 95 I have shown a key diagram indicating which direction of moment is to be taken as positive and which moment negative. In this case,

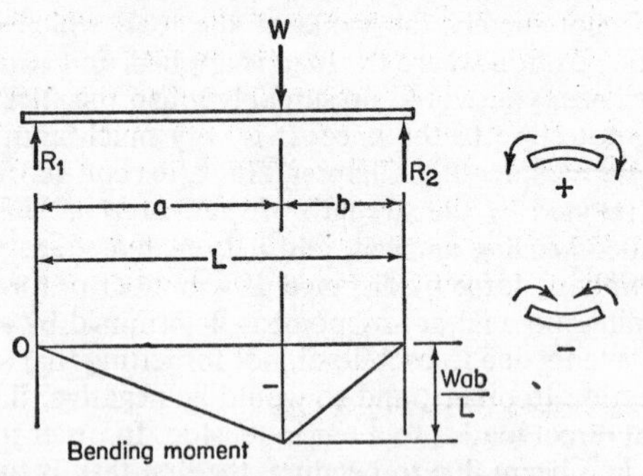

Fig. 95. Bending-moment diagram for a beam carrying a single load. Also shown is a key diagram defining positive and negative bending.

it can be seen that the moment is everywhere negative, and the moment diagram is accordingly drawn below the zero-line. If you think about the problem carefully, you should be able to reason out why it is not possible to determine the sign of the bending moment merely by its clockwise or anti-clockwise direction. This should become clear if you verify the above expression for the maximum bending moment by taking the moments of force to the right of the load-point, instead of to the left as I have done.

In the example given, if W is at the beam centre, $a = b = \frac{1}{2}L$ and the maximum bending moment becomes $\frac{1}{4}WL$. This is the case we shall shortly analyse on page 150.

If we have a load at the end of a horizontal beam which is fixed at the other end to a wall, the bending-moment diagram will be as shown in Fig. 96. This arrangement is called a cantilever. We have already encountered it in our examination of frameworks on page 101. It can easily be seen that the maximum bending moment is at the end supported by the wall, and has a value of WL. If the beam is going to break, this is the point where it will fail.

Many beams and girders carry loads which are not concentrated at a point but which are spread, often uniformly, along the beam. An example is a road bridge. Although the bridge carries concentrated loads, such as vehicles, a high proportion of the load carried is that due to its own weight. A uni-

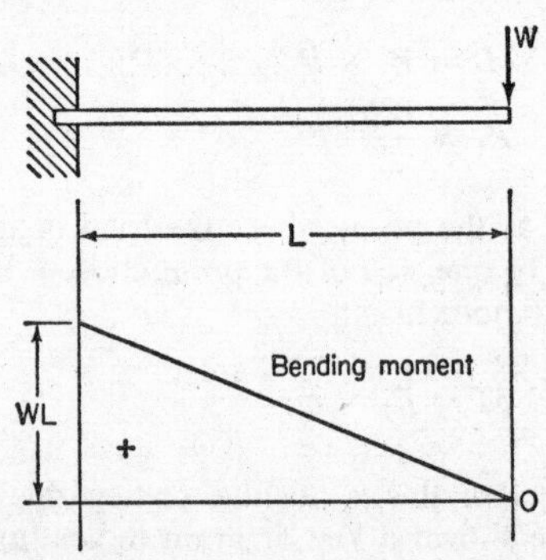

Fig. 96. The bending-moment diagram for a single load at the end of a cantilever.

formly distributed load on a beam supported at the two ends, together with the bending-moment diagram, is shown in Fig. 97.

It is seen that the graph of bending moment is now a curve, having a maximum value of $\frac{1}{8}WL$ at the beam centre—half the value it would have been had the load been all concentrated at the centre. The equation for this curve can be derived by examining the moments of forces to the left of a section a distance x from the left-hand end. Fig. 97 (*b*) shows these forces in the form of a free-body diagram. The forces consist of (i) the left-hand upward reaction,

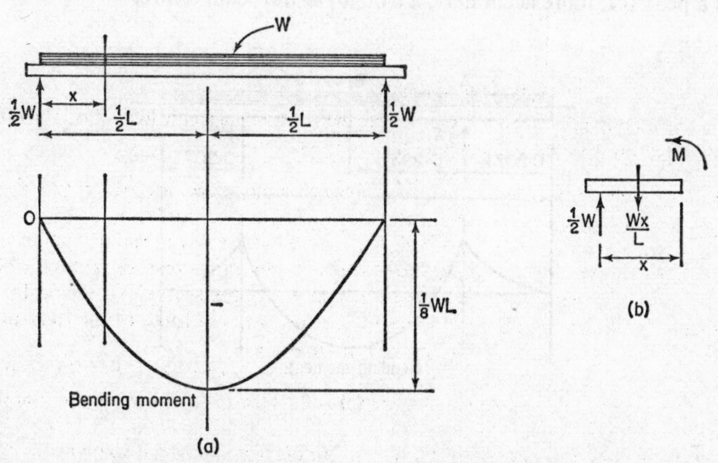

Fig. 97. The bending-moment diagram for a simply supported beam carrying a uniformly distributed load. At (*b*) is shown the free-body diagram for determining the bending moment at a distance x from the end.

which has a value of half the total load, and (ii) the downward weight of the portion of load of length x to the left of the point chosen. The magnitude of this load is $(x/L)W$, by simple proportion; its centre of gravity is half-way along it, a distance $\frac{1}{2}x$ from the point P. The bending moment at P, a distance x from the left-hand end, is therefore:

$$M = -(\tfrac{1}{2}W)x + \left(\frac{x}{L}W\right)(\tfrac{1}{2}x)$$

$$= \frac{1}{2}\frac{Wx^2}{L} - \frac{1}{2}Wx$$

This is the equation of a parabola (a type of curve we discussed in Chapter Two). When $x = \frac{1}{2}L$, at the centre of the beam:

$$M = \frac{1}{2}\left(\frac{W}{L}\right)(\tfrac{1}{2}L)^2 - \tfrac{1}{2}W(\tfrac{1}{2}L)$$

$$= \tfrac{1}{8}WL - \tfrac{1}{4}WL$$

$$= -\tfrac{1}{8}WL$$

as shown on the diagram.

You should be able to show fairly easily that a cantilever carrying a uniformly distributed load over a length L will have a maximum bending moment of $\frac{1}{2}WL$.

As a final, and revealing, example let us take the uniformly-loaded beam of Fig. 97 but, instead of supporting it at the two ends, we now move the supports inwards a distance of 0·207 L, where L is the total length of the beam. Fig. 98 shows the new arrangement, and it can be seen that the bending moment is positive over the overhanging portions, drops to zero, and thence to a peak (or, more accurately, a trough) at the beam centre.

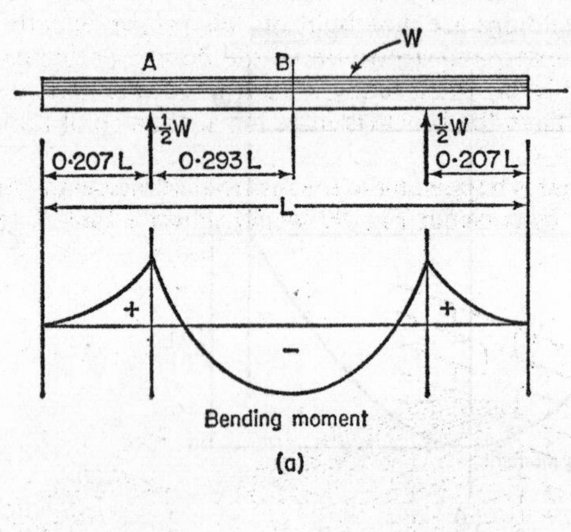

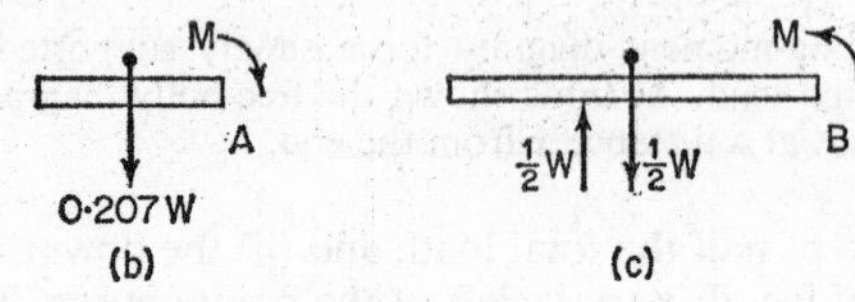

Fig. 98. (*a*) The bending-moment diagram for a beam with overhanging ends, and carrying a uniformly distributed load.
(*b*) The free-body diagram for calculating the bending moment at *A*.
(*c*) The free-body diagram for calculating the bending moment at *B*.

To evaluate the bending moment at the support-point A, the free-body diagram is shown in Fig. 98 (*b*).

$$\begin{aligned} M_A &= \text{Load} \times \text{Moment arm} \\ &= (0{\cdot}207W)(\tfrac{1}{2} \times 0{\cdot}207W) \\ &= +0{\cdot}021WL \end{aligned}$$

The bending moment at the centre B is evaluated using the free-body diagram of Fig. 98 (*c*).

$$\begin{aligned} M_B &= (0{\cdot}5W)(0{\cdot}25L) - (0{\cdot}5W)(0{\cdot}293L) \\ &= WL(0{\cdot}125 - 0{\cdot}146) \\ &= -0{\cdot}021WL \end{aligned}$$

The significance of the choice of support-points can now be seen. The maximum positive bending moment (at the supports) has the same value as the maximum negative bending moment (at the beam centre). But the further significance lies in the value itself as compared with the maximum value existing when the same loaded beam is supported at the two ends, as in Fig. 97. Assuming the load and the length of the beam to be the same in both cases, the maximum bending moment for the case of Fig. 97 was $0{\cdot}125WL$. In this present case, the maximum value is $0{\cdot}021WL$, approximately 17 per cent of the first value. It can be seen that this second method of supporting a uniform load will require beams of much smaller size than the first method. Many modern buildings are now built on this principle, with the main supporting columns inset, and a portion of the floors overhanging. At a more humble domestic level, when fixing a shelf to a wall on two brackets, it is good practice to have the brackets inset for a distance of about one-fifth of the shelf length.

Meanwhile, what is happening in the material of the beam itself? To answer this question, let us consider Fig. 99 which shows a timber beam fixed to a

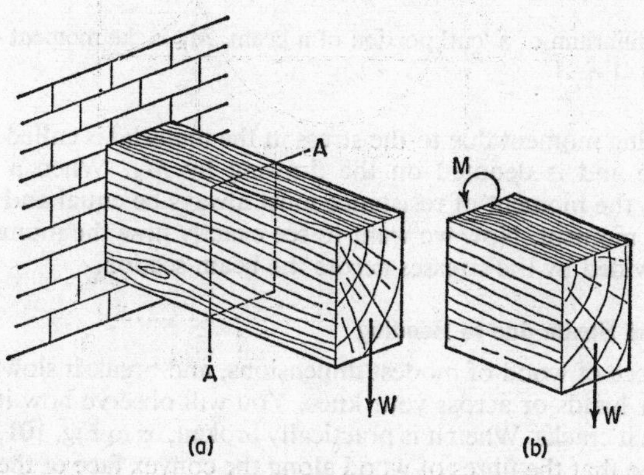

Fig. 99. A beam supporting a load. The section to the right of *A–A* is in equilibrium.

vertical wall at one end and supporting a downward load at the other. Our concern here is the stress caused by this load. As I showed earlier, to determine stress, the practice is to make an appropriate section through the loaded member and examine the forces on the section thus 'cut' away. This portion is shown in Fig. 99 (*b*).

We start with the obvious assertion that this isolated section of beam is in equilibrium—obvious because the section does not move up or down in response to any resultant force, so the forces on it must be in equilibrium. The only external force acting is the load *W*. What other forces are there? Accepting the assertion that forces are real tangible things, and have to be explained in terms of contact with real objects, the only remaining forces must be due to the contact of the 'cut' section with the parent beam. In simple words, the right-hand end of the beam is being held up by the left-hand end. (Similarly, the left-hand portion is being pulled down by the right-hand piece and is, in its turn, being held up by the contact with the wall.) The left-hand portion thus exerts forces on the right-hand portion through the medium of stress forces acting across the line of section *A–A*. These forces must be disposed in such a way that they set up a moment exactly equal to the bending moment of the load, but opposite in direction. In the present illustration, the external force W has a tendency to turn the beam clockwise. This must be opposed by the stress-forces, which therefore have an anti-clockwise sense. The result so far is illustrated by Fig. 100.

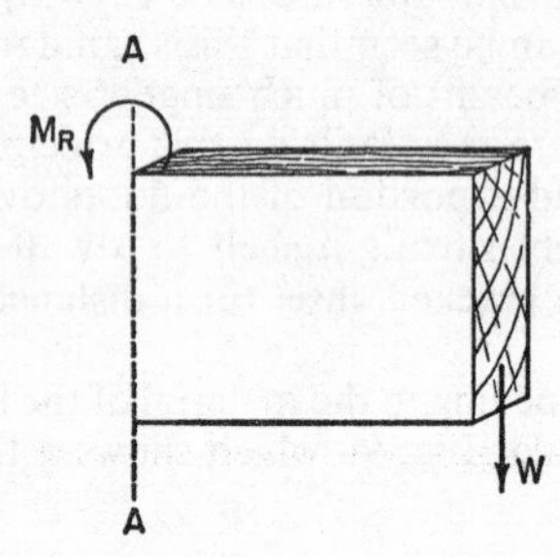

Fig. 100. Equilibrium of a 'cut' portion of a beam. M_R is the moment of resistance at the section at *A–A*.

The resisting moment due to the stress in the material is called the **moment of resistance** and is denoted on the drawing by M_R. When a beam is in equilibrium, the moment of resistance must always be equal and opposite to the bending moment. Now we want to see exactly how the moment of resistance is provided by the stresses across the beam section.

(2) Stress and Strain due to Bending

Take a piece of wood of modest dimensions, and break it slowly and carefully in your hands or across your knee. You will observe how it bends to a curve before it cracks. When it is practically broken, as in Fig. 101, you should be able to see that the fibres of wood along the convex face of the curve have been torn apart, while those along the concave face have been crushed together.

The convex face *a–a* in Fig. 101 is called the tension face, and the concave face *b–b* is the compression face. I imagine you begin to see already that we cannot apply the formula *stress* = *Load*/*Area* in such a case as this, because the stress is dependent upon which part of the cross-section we are examining.

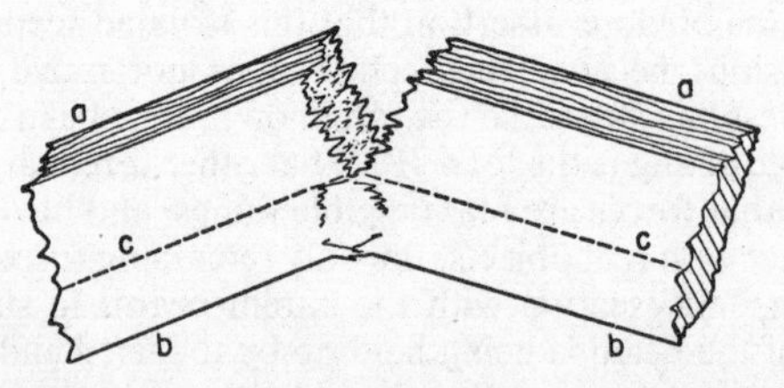

Fig. 101. A bending fracture. Stress along the face *a–a* is tensile, that along the face *b–b* is compressive; *c–c* is the neutral plane.

We have a typical example of non-uniformity of stress, such as we discussed in Chapter Eight. Not only does the stress vary, it changes from tensile to compressive.

Since the stress is tensile at one face of the beam and compressive at the other, we may conclude there must be a point somewhere across the cross-section where stress is neither tensile nor compressive, but zero. Such a point is along the line *c–c*, and we call *c–c* the **neutral plane** for the beam.

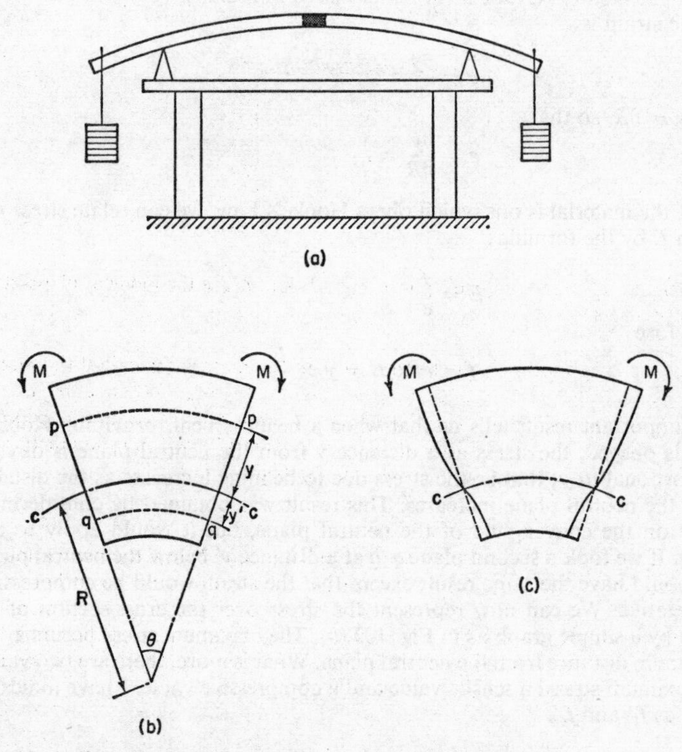

Fig. 102. A diagram of bending strain; *c–c* is the neutral plane.

We shall now look closely at a portion of a bent beam and see exactly what is happening at various points. Fig. 102 (*a*) shows a simple beam bent by applying a twist at the ends. We shall study what happens to the small shaded portion of the beam, which is shown very much enlarged in Fig. 102 (*b*). Suppose that before it was bent, it was of length *x*. Bending causes it to take up the curved shape shown, and we shall call the radius of this curve *R*. *c–c* is the neutral plane, i.e. the plane of zero stress and therefore of zero strain. The curved length of the line *c–c* must still be *x* because there is no strain along this line.

Look at the line *p–p*, which is a plane of material a distance *y* from *c–c*. Since the portion of beam was rectangular before bending, the original length of *p–p* must have been *x*. You can see from the diagram that it is now larger. It has been subjected to strain, which we can call *e*. The increase of length of *p–p* is the difference between its length on the diagram (its strained length) and its original length *x*. This increase is given by the simple calculation:

$$\text{Increase} = \theta(R + y) - \theta R$$
$$= \theta y$$

where the angle θ (Greek theta) is measured in radians.
So the strain is:

$$e = \frac{\theta y}{x}$$

But $x = \theta R$, so that:

$$e = \frac{\theta y}{\theta R} = \frac{y}{R}$$

But if the material is one which obeys Hooke's Law, we can relate stress f to strain E by the formula:

$$E = \frac{f}{e}$$

Therefore

$$f = e \times E = y \times \frac{E}{R}$$

This important result tells us that when a beam is bent, providing Hooke's Law is obeyed, the stress at a distance y from the neutral plane is directly proportional to y; that is, the stress due to bending increases as the distance from the neutral plane increases. This result was obtained by considering a plane on the convex side of the neutral plane, but it would apply to any plane. If we took a second plane *q–q* at a distance y' below the neutral plane, we should have the same result except that the strain would be compressive, or negative. We can now represent the stress over the cross-section of the beam by a simple graph as in Fig. 103 (*b*). The maximum stress occurs at the maximum distance from the neutral plane. What is more, there are two values of maximum stress: a tensile value and a compressive value. I have indicated these as f_t' and f_c'.

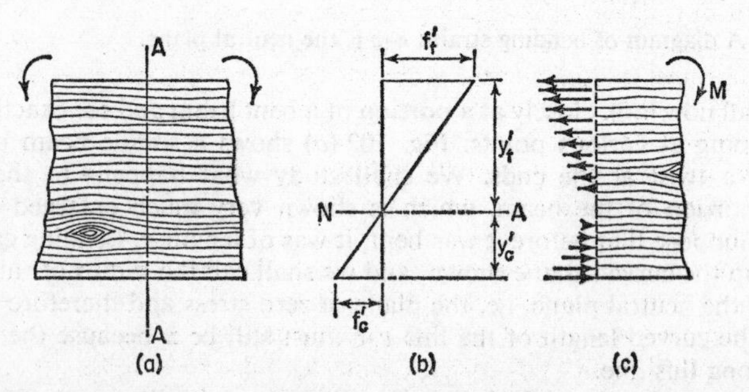

Fig. 103. Stress variation over a beam cross-section: (*a*) section *A–A* across the beam; (*b*) graph of stress variation; (*c*) forces over the 'cut' section.

Before going on to evaluate the stress, it is only fair to point out that our formula $f = y \times E/R$ was obtained by making certain simplifying assumptions which I did not state at the time. You can perceive how the whole argument depends on the evaluation of strain based on the sector of a circle, which Fig. 102 illustrates. But of course, we have no guarantee that our beam is conveniently going to bend in this fashion to suit our formula. For instance, suppose the original rectangular portion we considered bent into some such form indicated by Fig. 102 (*c*). Then our evaluation of strain is completely invalid. We also assume a perfect material, i.e. one having no variation of quality or of elasticity throughout. In actual fact, these assumptions are reasonably accurate, providing the amount of bending is small.

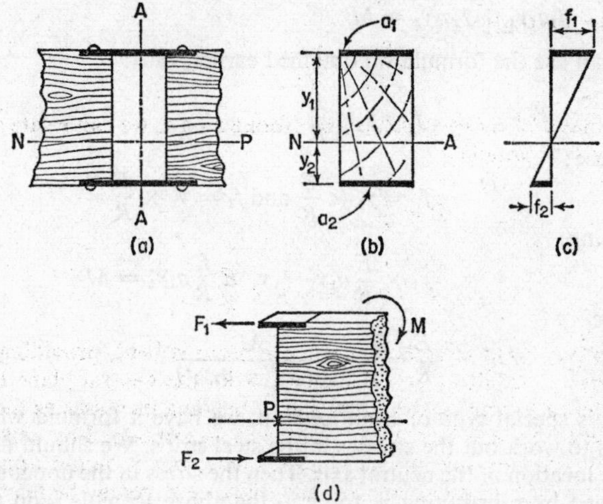

Fig. 104. A special beam with tension and compression elements: (*a*) section of the beam; (*b*) end-view of the section; (*c*) variation of stress; (*d*) the free-body diagram.

(3) Bending Stress

In order to obtain the values of maximum stress in terms of the load on the beam, we have to write an equation of equilibrium of the isolated portion of beam, not of forces, but of moment. If we consider the forces acting on the end of the portion, as I have done in Fig. 103 (*c*), what we have is an infinite number of forces acting over the whole of the end of the portion and getting progressively smaller and smaller as they approach the neutral plane. This is really a problem for solution by the calculus, but we can form some idea of the answer by imagining first of all a special type of beam shown in Fig. 104.

In the relevant portion of the beam the timber has been removed and replaced by two thin steel strips, one at the top and one at the bottom. We can imagine these strips to be extremely thin, the advantage being that we can then assume that the stress across them does not vary. The cross-sectional areas of the steel are a_1 and a_2, and their centres are respectively y_1 and y_2 from the neutral plane. Fig. 104 (*b*) shows an end-elevation of the portion. The line where the neutral plane cuts this cross-section is called the **neutral axis** and is shown as *N–A*. Fig. 104 (*c*) shows the graph of variation of stress across the section, but since there is no material over the section except the thin strips, it is only the blacked-in part of the graph that has interest for us. The *forces* in the two steel strips we can call F_1 and F_2. The free-body diagram is shown in Fig. 104 (*d*) with these forces acting respectively to left and right, and the bending moment of the external forces shown merely by M. Writing an equation of equilibrium of moments about the point P:

$$F_1 \times y_1 + F_2 \times y_2 = M$$

bearing in mind that M is already a moment, and so should not be multiplied by a distance.

Replacing forces by (stress) × (area):

$$f_1 a_1 y_1 + f_2 a_2 y_2 = M$$

But we can use the formula we obtained earlier, thus:

$$f = y \times \frac{E}{R}$$

In this case:

$$f_1 = y_1 \times \frac{E}{R} \text{ and } f_2 = y_2 \times \frac{E}{R}$$

Substituting:

$$y_1 \times \frac{E}{R} a_1 y_1 + y_2 \times \frac{E}{R} a_2 y_2 = M$$

Therefore

$$\frac{E}{R} = \frac{f}{y} = \frac{M}{(a_1 y_1^2 + a_2 y_2^2)}$$

So for this special type of beam section, we have a formula which would enable us to work out the stresses in the steel strips. We should first have to know the location of the neutral axis. Then the stress in the upper strip would be obtained by substituting y_1 for y in the above formula, and that in the lower by substituting y_2 for y.

If we now introduce a further complication by adding a third strip at the section, a distance y_3 from the neutral axis, the free-body diagram will have a third force F_3 as shown in Fig. 105.

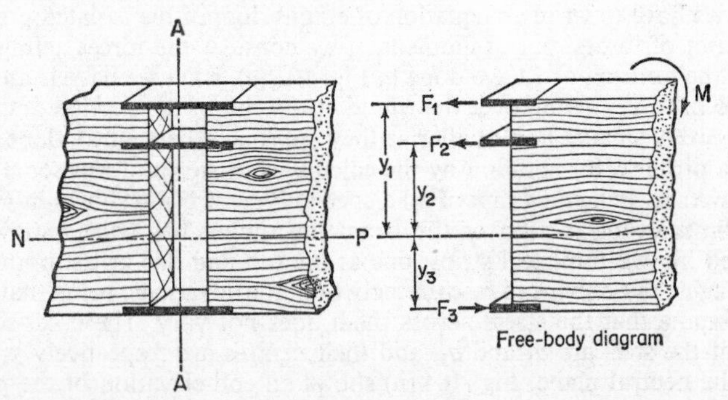

Fig. 105. Three load-bearing elements across the beam cross-section. Note that F_3 is compressive because it is below the neutral plane.

The equilibrium equation will now be:

$$F_1 \times y_1 + F_2 \times y_2 + F_3 \times y_3 = M$$

and skipping the detailed analysis, which is exactly as before, the result will be:

$$\frac{E}{R} = \frac{f}{y} = \frac{M}{(a_1y_1^2 + a_2y_2^2 + a_3y_3^2)}$$

When we come to consider a section across the ordinary beam without the steel strips, we can imagine the section to be divided into strips, each providing its own increment of force to supply the moment of resistance of the section, as shown in Fig. 106.

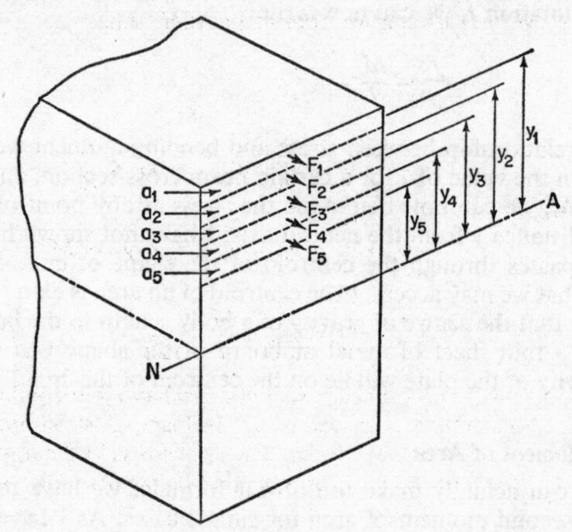

Fig. 106. A continuous cross-section may be imagined as being made up of a number of separate strips.

The same sort of analysis as before would give:

$$\frac{E}{R} = \frac{f}{y} = \frac{M}{(a_1y_1^2 + a_2y_2^2 + a_3y_3^2 + \ldots)}$$

the bracketed term containing as many terms as there were strips. This bracketed term is very similar to the term we obtained on page 61 for the moment of inertia of a spinning body. In that case, the terms represented the manner in which the mass was distributed about the axis. Here, they represent the way in which the *area* of the cross-section is distributed about the neutral axis. We call this term the **second moment of area** of the section. It is a geometrical property of the shape and area of the section, and can be evaluated for any particular case. For simple regular figures such as rectangles or circles, the value can be determined by using the calculus. For less regular figures, a geometrical method would have to be used, based generally on the division of the areas into strips.

Due to the mathematical similarity between this quantity and the moment of inertia, the second moment of area has in the past been called the moment of inertia, and there may be books still in circulation in which the term is still used. A legacy from the past is the use of the symbol I for this quantity—a notation which has steadily resisted attempts to change it. Students are occasionally prone to confuse the two terms, although to do so is as logical as to confuse a mass with an area. If you are one of those unfortunate people, in all probability you have either failed to understand the treatment of the subject, or, in following the mathematical analysis, you have lost sight of the physical situation and have become hypnotized by formulae—a far from uncommon state among students. I can only remind you that, as far as this and many other subjects are concerned, mathematics exists to solve physical problems, not to replace them.

Using the notation I, we can now write:

$$\frac{f}{y}=\frac{M}{I}$$

which is the relationship between stress and bending moment we have been seeking. Given the value of I for a certain beam cross-section, and the bending moment M, we can now determine the stress at any point on the cross-section at a distance y from the neutral axis. I have not shown here that the neutral axis passes through the centroid of the shape of cross-section, but this is a fact that we may accept. (The centroid of an area is akin to an area in the same way that the centre of gravity of a body is akin to the body. In fact, if you make a thin sheet of metal or board of the shape you require, the centre of gravity of the plate will lie on the centroid of the area.)

(4) Second Moment of Area

Before we can actually make use of our formula, we have to be able to evaluate the second moment of area for simple cases. As I have stated, the calculus is used to arrive at the values for regular shapes, and we shall use the results. For shapes which are not simple, values are obtained by other means, usually graphical, and results of all standard sections used by beam designers are provided in standard tables.

For the solid rectangular section shown in Fig. 107 (*a*) the value of the second moment of area is given by:

$$I=\frac{BD^3}{12}$$

The neutral axis passes through the centre of symmetry of the area. For the three 'hollow' sections of Fig. 107 (*b*) the value is:

$$I=\frac{BD^3}{12}-\frac{bd^3}{12}$$

This result is arrived at, in effect, by subtracting the 'cut-away' portion of the section from the overall rectangular section. Again the neutral axis passes through the centre of symmetry.

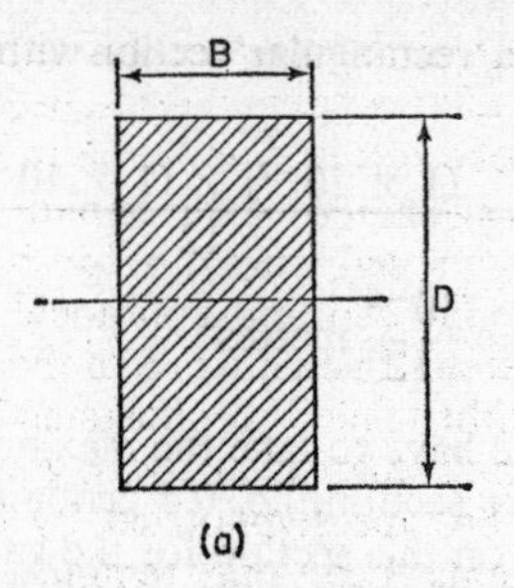

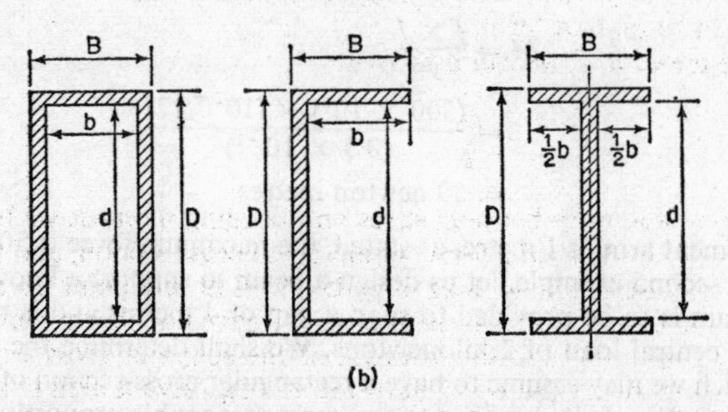

Fig. 107. (*a*) A solid rectangular section. (*b*) Hollow rectangular sections.

For the hollow circular section of Fig. 108 the value is:

$$I = \frac{\pi}{64}(D^4 - d^4)$$

and again the neutral axis passes through the centre of symmetry of the section.

We are now in a position to tackle a few simple calculations. For the first one, let us go back to our original example of the square-section bar of area 1 square centimetre at the beginning of this chapter. I stated that such a bar would withstand a load of approximately 50 newtons if held in a vice with the load applied transversely 1 metre from the vice. Let me now show how this figure was obtained. We have first to decide on a value of 'dangerously high' stress in the material, and a reasonable estimate for this would be 300 meganewtons per square metre (approximately 20 tons per square inch). Next we

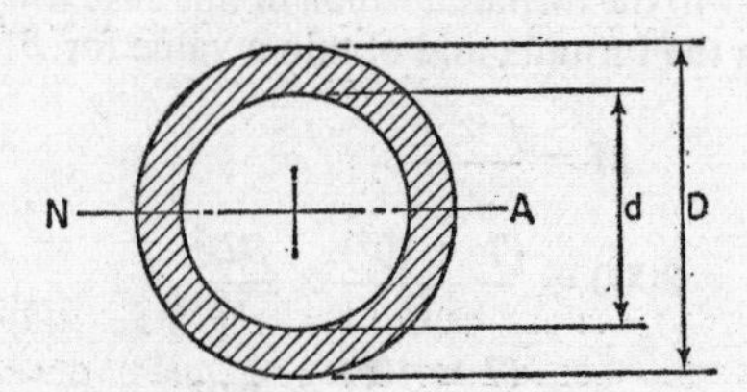

Fig. 108. A hollow circular section.

can evaluate I. The bar has a 'rectangular' section with $B = D = 1$ centimetre; so:

$$I = \frac{(1 \times 10^{-2}) \times (1 \times 10^{-2})^3}{12}$$

$$= \frac{10^{-8}}{12} \text{ metres}^4$$

For the maximum stress, we have to take the maximum value of y, which is half the depth of the square section, i.e. 0·5 centimetres. So the maximum allowable bending moment on this section for the value of stress assumed is given by:

$$M = \frac{f \times I}{y}$$

$$= \frac{(300 \times 10^6) \times (10^{-8}/12)}{(0{\cdot}5 \times 10^{-2})}$$

$$= 50 \text{ newton metres}$$

If the 'moment arm' is 1 metre, as stated, the maximum force is 50 newtons.

For the second example, let us design a beam to support a known load. A timber beam is to be provided to span a gap of 4 metres and is required to support a central load of 2 kilonewtons. We shall determine the size of the beam, which we may assume to have a rectangular cross-section of breadth B and depth D. We shall also have to assume a reasonable proportion of cross-section for the beam. Let us take the depth D to be three times the breadth. Finally, let us take the maximum permissible stress in the timber to be 7 meganewtons per square metre (approximately 1000 pounds per square inch).

The arrangement is shown in Fig. 109 (*a*). The free-body diagram in Fig. 109 (*c*) discloses the two upward reaction forces at the support points, each of which must be equal to half the central load. We first have to decide where along the beam the bending moment will be at its maximum possible value. If we examine the moment at a section A–A a distance x from the right-hand end, the moment will be the product of force on this end (1000 newtons) and distance x from section. For the moment to be a maximum, x must be as large as possible, in this case 2 metres. If we make it any larger, the section will then include the downward central load which will have an opposite moment to the reaction. So the maximum bending moment on the beam is at the centre, under the load, and its value is (1000 × 2) = 2000 newton metres.

Fig. 109 (*b*) shows the cross-section of the beam. The maximum stress is known to occur at the maximum distance from the neutral axis, i.e. the maximum value of y in the formula, which in this case will be $\frac{1}{2}D$. We can now substitute values in the formula and obtain a value for B:

$$M = \frac{f \times I}{y}$$

$$2000 = \frac{(7 \times 10^6)}{\frac{1}{2}D} \times \frac{BD^3}{12}$$

$$= \frac{(7 \times 10^6)}{\frac{1}{2} \times 12} \times BD^2$$

$$= \frac{7 \times 10^6}{6} \times B \times (3B)^2$$

$$= \frac{7 \times 10^6}{6} \times 9 \times B^3$$

$$B^3 = \frac{2000 \times 6}{7 \times 10^6 \times 9} = 0{\cdot}19 \times 10^{-3}$$

Therefore $B = 0{\cdot}575 \times 10^{-1}$ metres $= 5{\cdot}75$ centimetres

and the depth of the beam will be three times this value, namely 17·25 centimetres.

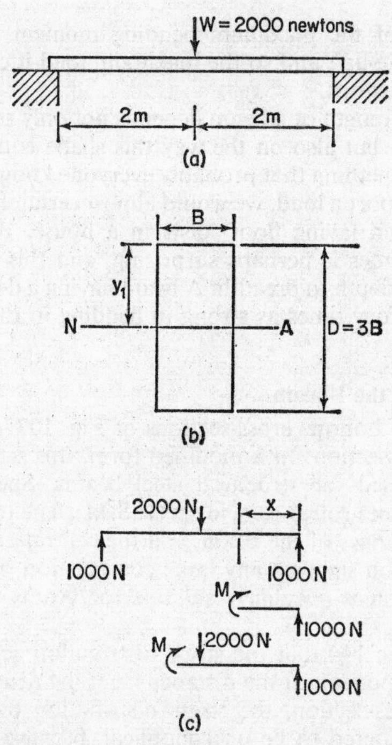

Fig. 109. A centrally-loaded beam: (*a*) loading arrangement; (*b*) the cross-section; (*c*) the free-body diagram.

For a revealing comparison, let us take a beam of these dimensions and see how strong it would be if we turned the section through 90° so that the long side of the rectangle is now horizontal. Then $B = 17{\cdot}25$ cm and $D = 5{\cdot}75$ cm. The second moment of area is now:

$$I = \frac{BD^3}{12}$$

$$= \frac{17{\cdot}25 \times 5{\cdot}75^3}{12}$$

$$= 274 \text{ centimetres}^4$$
$$= 274 \times 10^{-8} \text{ metres}^4$$

y is half the new depth, i.e. $\frac{1}{2} \times 5{\cdot}75$ centimetres. So the maximum bending moment is:

$$M = \frac{f \times I}{y}$$
$$= \frac{(7 \times 10^6) \times (274 \times 10^{-8})}{\frac{1}{2} \times 5{\cdot}75 \times 10^{-2}}$$
$$= 667 \text{ newton metres}$$

This is one-third of the maximum bending moment permitted when the beam is placed 'edge-up', and so the maximum load it could carry would be $\frac{1}{3} \times 2000 = 667$ newtons.

So the bending strength of a beam depends not only on the area and shape of the cross-section but also on the way this shape is disposed to resist the moment. This is something that probably everyone knows. If we were using a timber beam to support a load, we would almost certainly stand it on edge, as a builder does when laying floor joists in a house. But multiplication of strength of three times is perhaps surprising, and this factor is greater the greater the ratio of depth to breadth. A beam having a depth of four times the breadth would be four times as strong in bending in the 'edge-up' position, and so on.

(5) The Function of the I-beam

The third of the 'hollow' cross-sections of Fig. 107 (*b*) is, for an obvious reason, called an I-section. In a modified form, this section is probably the most commonly used for structural steel beams. Such beams are called I-beams, or sometimes rolled-steel joists (R.S.J.s). The top and bottom of the 'I' are called the flanges of the beam, and the central portion the web. Such beams are a common sight on any large construction site. Let us try to find out why this section is peculiarly suitable for beams which have to resist bending moment.

I showed on page 144 that the stress distribution across the section of a beam varies in proportion to the distance from the neutral axis. For a beam of rectangular cross-section, the stress distribution over the depth of the beam can be considered to be uneconomical, because the maximum stress occurs at only two points on the cross-section: at the upper and the lower face. The remainder of the material is stressed to a value below its permissible limit. By this standard, the rectangular cross-section is wasteful of material. This does not matter for a relatively cheap material such as timber (which cannot be formed into convenient shapes anyway), but for steel, which is expensive, and for which a relatively significant fraction of the load of a beam is that due to its own weight, it becomes necessary to dispose the material more economically. The I-section is thus designed to dispose the material of the beam, or at least a large proportion of it, at *the points where the stress is greatest*, i.e. near the upper and lower surfaces. The result is that most of the area of cross-section will function under load at or near its maximum permissible stress-value.

Another way of looking at the I-section beam is to regard it as a rectangular section with most of the under-stressed material removed. Such a beam is *weaker* than the corresponding rectangular section from which it is derived, but it is many times *lighter*, so the strength-to-weight ratio is increased. As an example, a beam of I-section $5\frac{1}{4}$ inches wide and 8 inches deep has a value for I of 69·2 inches^4. (This value is taken from a standard Tables of Sections.) The value of I for a rectangle of the same overall dimensions is $1/12 \times 5\frac{1}{4} \times 8^3 = 224$ inches^4, which is 3·24 times as much. But the cross-sectional area of the I-section is only 5·88 square inches as compared with 42 square inches for the rectangle, which is 7·14 times as much. A simple calculation shows that 72 per cent of the area of cross-section of the I-beam lies more than 3·35 inches from the neutral axis, so the stress over this 72 per cent would be of greater value than 84 per cent of its maximum permissible value.

CHAPTER TEN

STRESSES DUE TO TORSION

In this chapter, we shall examine the stresses set up when a torque or twist is applied to a shaft. We shall find, rather surprisingly perhaps, that the analysis is very closely allied to that of the previous chapter, and that the resulting formula relating torque, stress and deformation is very closely allied to the formula relating bending moment, stress and deformation. However, whereas the cross-section of a beam can vary widely in area and shape, the cross-section of a shaft is almost invariably circular, and we shall limit our investigation to that particular case. We shall find in consequence that it is much easier to determine the deformation of a twisted shaft than to find the deflection of a bent beam.

(1) Stress Due to Torsion

As with bending, the first thing to determine is the nature of the stress caused by a torque, or twist, on a shaft. As with the bending of a beam, the method of solution consists of making an imaginary section across a loaded shaft, and examining the isolated portion as a free body with a view to writing an equation of equilibrium. However, when we write such an equation, we must remember that the torque T is essentially a moment, i.e. a product of force and distance.

Fig. 110 shows the shaft, with the 'cut' section drawn separately as a free body. The torque is resisted by the stress-force over the 'cut' face. The torque is really a moment of force about the shaft axis x–y, as distinct from the beam of the previous chapter, where the moment was about an axis perpendicular to the length of the beam. So the equal and opposite stress-force must act in the same sense, i.e. it must exert a moment about the shaft axis—although opposite in direction to the torque. The stress must therefore be *parallel* to the plane of section, unlike the stress due to bending, which was found to be per-

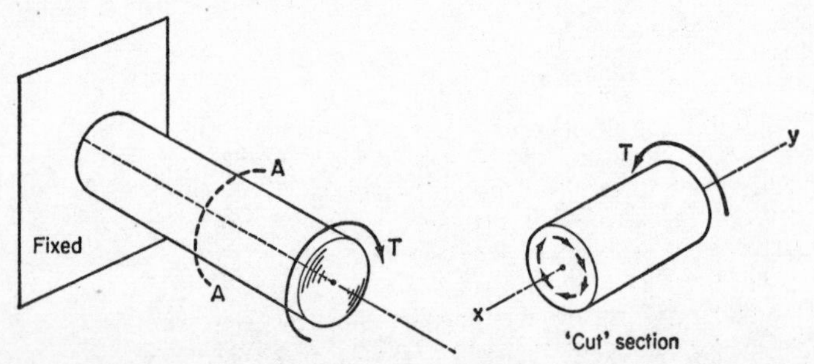

Fig. 110. Equilibrium of a 'cut' section of shaft. The external torque, or twist, is balanced by shear stress forces over the 'cut' section.

pendicular to the plane of section. Stresses acting parallel to a plane are shear stresses, and as I foretold in Chapter Eight, a torque on a shaft gives rise to a pure shear stress in the material of the shaft.

(2) Shear Strain in a Shaft

Let us now consider the deformation of the shaft. Fig. 111 shows a diagram of a circular shaft which is to be imagined as fixed at the left-hand end and subjected to a torque T at the right-hand end. The result is that the free end twists through an angle which I have shown in the diagram as θ. I am

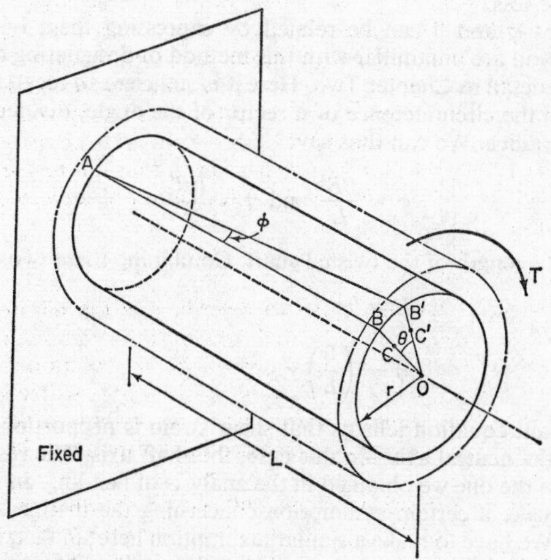

Fig. 111. Diagram of strain in a shaft in torsion.

going to obtain an expression for the shear strain along this shaft for a cylinder of radius r within the shaft; note that r is *not* the outside radius of the shaft. For clarity, this cylinder is shown full, the remainder of the shaft being indicated in faint outline. Imagine that, before twist takes place, a line is drawn along this cylinder from a point A at the fixed end to a point B at the free end, the line being parallel to the shaft axis. Twisting will cause the point B to move to B'. Since A remains fixed, the line AB will become the line AB', which will now not be parallel to the shaft axis. The angle between the line AB' and the original line AB is shown as φ (phi).

It should be fairly clear, if you have followed Section 10 of Chapter 8, that this angle φ is the shear strain at the radius r. The difference here is that the strain takes place in a rotary mode, instead of a linear one. To illustrate shear strain in a linear mode, I used the notion of the leaves of a book. If you now remove the binding from the book so that you are just left with the separate leaves, and if, instead of pushing to one side with your hand, you press downwards and twist, then the pages should all slip round a small amount and the pile should take up a spiral formation. If, finally, you imagine circular leaves, instead of rectangular, you have a fair model of a shaft in torsion.

In analysing bending, we came upon the idea of a neutral plane, i.e. a plane in the material where no stress existed. In analysing torsion, the shaft axis itself is analogous to the neutral plane. There can be no strain along the shaft axis, and it is the one region of the shaft which suffers no deformation.

How does the strain vary with the radius? Remember that φ is the strain at a particular radius r. At this radius, B moves to B' when the shaft twists through an angle θ. If we had taken a smaller radius such as OC, then C would have moved to C'. The angle of twist θ is the same, but the angle of strain will be less.

The angles φ and θ can be related by expressing them both in radian measure. If you are unfamiliar with this method of designating an angle, it is discussed in detail in Chapter Two. Here it is sufficient to recall that an angle in radians is the circumference of a sector of the angle, divided by the corresponding radius. We can thus say:

$$\varphi = \frac{BB'}{L} \text{ and } \theta = \frac{BB'}{r}$$

where L is the length of the twisted shaft. Combining these two equations:

$$\varphi L = \theta r$$

Therefore

$$\varphi = \left(\frac{\theta}{L}\right) r$$

This important equation tells us that shear strain is proportional to the distance from the neutral axis—in this case, the shaft axis. The result is exactly analogous to the one we obtained in the analysis of bending. In that analysis, we had to make a certain assumption concerning the manner in which the beam bent. We have to make a similar assumption here; in fact, it has already been made. In discussing the deformation at the smaller radius, I assumed that both B' and C' would still lie on the same radius when the shaft was twisted, and of course we have no guarantee that the shaft will be so obliging as to conform to this requirement. Physically, the assumption means that an original radial line will remain radial when the shaft is twisted and, further, that it will remain in the same plane perpendicular to the shaft axis. Or, to use the jargon of stress-analysis, 'plane sections of the shaft remain plane sections under torsion'. It is sufficient to add that, for circular shafts, when the strain is small, the assumption is justified by observation. But the twisting of shafts of square or rectangular cross-section does not conform to this ideal, and a different analysis is necessary.

So far, we have related the shear strain to the angle of twist. We can now use the formula from page 132:

$$\frac{q}{\varphi} = G$$

and eliminating φ from this and the previous equation:

$$\frac{q}{G} = \frac{\theta r}{L}$$

Comparison with the equation of bending on page 144 shows that the terms are in all respects analogous, if we replace the radius of curvature R by the terms L/θ.

(3) Torque, Stress and Deformation

The formula we have derived relates stress q to deformation θ. We now need to relate both these quantities to the applied torque T. As with bending of a beam, this problem is made simpler by first picturing a special shaft having a portion made of a very thin cylinder, which we may call a 'torsion element'.

Fig. 112 shows such a shaft. Portion A is connected to portion B by this single thin-walled cylinder, which we will give a mean radius r_1 and a cross-sectional

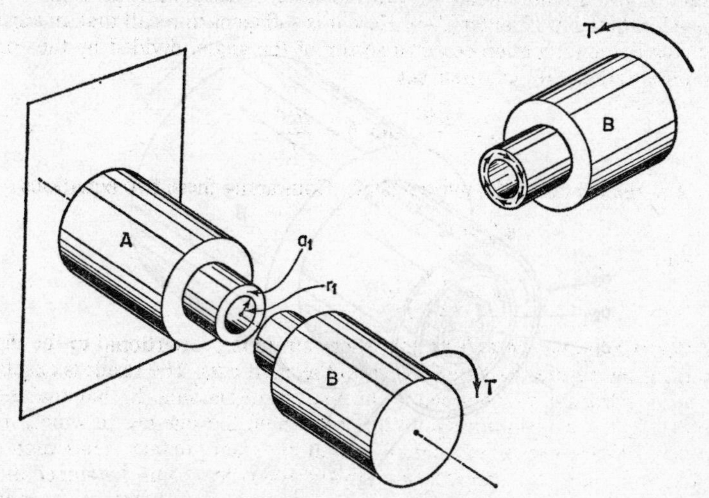

Fig. 112. A special shaft with a single torsion element.

area a_1. A is the fixed part of the shaft, and torque T is applied to B as shown, so that the full value of the torque is transmitted by the torsion element. To find the stress on this element, we make a section through it and examine the portion B as a free-body. The free-body diagram of B is shown separately. We know that the stress around the annular 'cut' face is a shear stress, running around the face, as indicated by the small arrows. Because of the thinness of the element, the variation of radius can be neglected, and the stress can thus be assumed constant over the cylinder wall. Calling this stress q_1, the force due to it will be $q_1 a_1$. The torque exerted by this stress-force, which we can call the 'torque of resistance', will be the product of force and effective radius, that is:

$$T = q_1 a_1 r_1$$

from which we obtain

$$q_1 = \frac{T}{a_1 r_1}$$

But we have already shown that

$$\frac{q_1}{G} = \frac{\theta r_1}{L}$$

Therefore

$$\frac{G\theta}{L} = \frac{q_1}{r_1} = \frac{T}{a_1 r_1^2}$$

This formula would enable us to calculate the value of the stress q_1 in the wall of the torsion element.

Now imagine the two portions of shaft connected by, say, three thin cylinders or torsion elements, each of a different radius. The new free-body

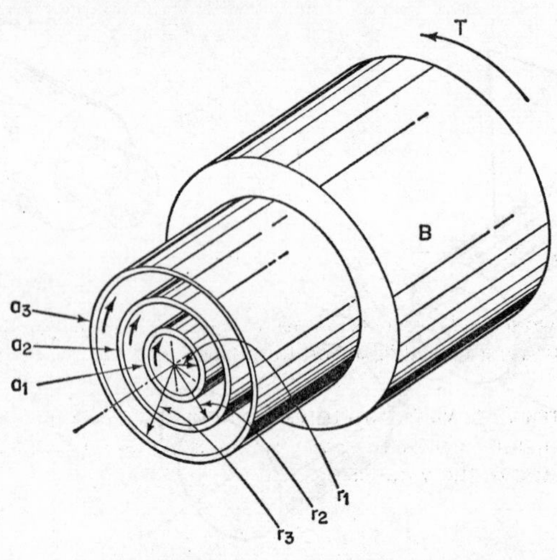

Fig. 113. The free-body diagram for a section of shaft with three torsion elements.

diagram of portion *B* is shown in Fig. 113. The torque at *B* is now resisted by stress forces at the three radii r_1, r_2 and r_3. Calling the respective stresses q_1, q_2 and q_3,

$$T = q_1 a_1 r_1 + q_2 a_2 r_2 + q_3 a_3 r_3$$

and substituting for the values of q:

$$T = \left(\frac{G\theta}{L} r_1\right) a_1 r_1 + \left(\frac{G\theta}{L} r_2\right) a_2 r_2 + \left(\frac{G\theta}{L} r_3\right) a_3 r_3$$

Therefore

$$\frac{G\theta}{L} = \frac{q}{r} = \frac{T}{(a_1 r_1^2 + a_2 r_2^2 + a_3 r_3^2)}$$

From this last formula we can calculate q_1, q_2 and q_3 by substituting the appropriate value of radius in the middle term.

For a continuous shaft, i.e. a shaft not made up of separate torsion elements, we have in effect an infinite number of elements joining the sections A and B. The collection of terms in the bracket is analogous to the second moment of area of a section in bending: in fact, it *is* a second moment of area. But its value will not be the same as that obtained for the bending case, because the geometrical summation of the elements is about a different axis. I have tried to show the difference in Fig. 114.

In direct bending, we considered thin strips which were at a constant distance from the neutral axis, as shown in Fig. 114 (*a*), the neutral axis being in the plane of the shaft section. The torsion element we have considered here is at a constant distance from the *shaft* axis, which is perpendicular to the plane of section. The quantity $(a_1r_1^2 + a_2r_2^2 + a_3r_3^2 + \ldots)$ is called the **polar second moment of area**, and is denoted by J. Like the ordinary second

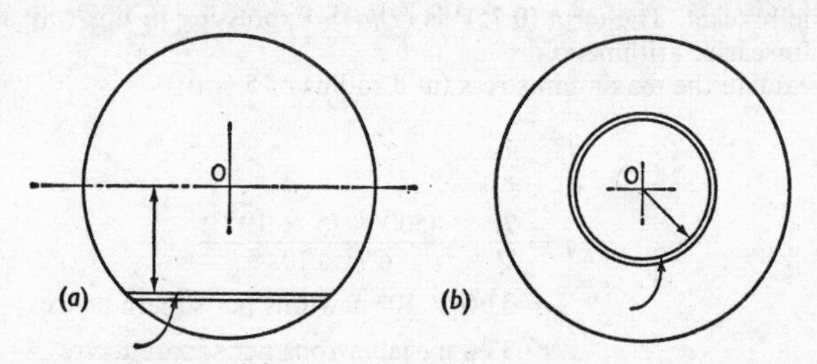

Fig. 114. The two second moments of area of a circular section: (*a*) a typical 'element' for a bent shaft; (*b*) a typical 'element' for a twisted shaft.

moment of area I, its value can be easily obtained for simple sections by the use of the calculus. For a hollow circular shaft having outside and inside diameters D and d, the value is:

$$J = \frac{\pi}{32}(D^4 - d^4)$$

which is seen to be twice the value of I for the same section. The transmission of torque is almost always by shafts of circular cross-section, and so this is the only value of J we are concerned with.

The final formula relating torque, stress and deformation is thus:

$$\frac{T}{J} = \frac{q}{r} = \frac{G\theta}{L}$$

When applying this formula, we must remember that q is the stress at *any* radius r. However, we are almost always concerned with the greatest shear stress in the shaft material, and this occurs at the *greatest* radius.

(4) Some Examples

For our first example we want to determine the stress and twist for a given value of torque. We shall assume a hollow shaft, of outside diameter 10 centimetres and inside diameter 7·5 cm. Let us calculate the stress and the angle of twist over a length of 1 metre when the shaft transmits a torque of 4500 newton metres.

When the shaft size is known, the first thing is to evaluate J. Using the formula:

$$J = \frac{\pi}{32}(D^4 - d^4)$$

$$= \frac{\pi}{32} D^4(1 - (0{\cdot}75)^4)$$

$$= \frac{\pi}{32} \times 10^4 \times 10^{-8}(1 - 0{\cdot}317)$$

$$= 6{\cdot}7 \times 10^{-6} \text{ metres}^4$$

Note the conversion of all dimensions to metres, and the factorizing of D^4 from the bracket. The term $(0{\cdot}75)^4$ is $(D/d)^4$. Expressing in this way always makes for easier arithmetic.

To evaluate the maximum stress (at a radius of 5 cm):

$$\frac{T}{J} = \frac{q}{r}$$

$$q = \frac{Tr}{J} = \frac{4500 \times (5 \times 10^{-2})}{6{\cdot}7 \times 10^{-6}}$$

$$= 33{\cdot}6 \times 10^6 \text{ newtons per square metre}$$

$$= 33{\cdot}6 \text{ meganewtons per square metre}$$

$$= 5000 \text{ pounds per square inch approximately}$$

The *least* stress in the shaft will be at the inner radius and so will be three-quarters of the above value.

To determine the angle of twist, we can equate T/J or q/r to $G\theta/L$

$$\frac{T}{J} = \frac{G\theta}{L}$$

$$\theta = \frac{TL}{GJ} = \frac{4500 \times 1}{(80 \times 10^9) \times (6{\cdot}7 \times 10^{-6})}$$

$$= 0{\cdot}0084 \text{ radians}$$

$$= 0{\cdot}0084 \times \frac{360}{2\pi} \text{ degrees}$$

$$= 0{\cdot}48^\circ$$

In the above calculation, the value of G of 80 giganewtons per square metre stated on page 132 is used.

As as second example, let us work the other way and determine the dimensions of a shaft required to transmit a torque of 50 kilonewton metres. This is a sizeable torque, such as may be encountered in a marine propeller drive. We shall take two possible cases: first, a solid shaft; secondly, a hollow shaft having a bore diameter of 0·9 times the outside diameter. We shall take the same value of maximum permissible stress, and this can be 30 meganewtons per square metre.

We use the formula $T/J = q/r$. Both J and r are unknown, but both may be expressed in terms of the unknown diameter D. For the solid shaft this gives:

$$\frac{(50 \times 10^3)}{(\pi/32)D^4} = \frac{(30 \times 10^6)}{\frac{1}{2}D}$$

Rearranging:

$$D^3 = \frac{50 \times 10^3 \times 32 \times \frac{1}{2}}{\pi \times 30 \times 10^6}$$
$$= 8{\cdot}5 \times 10^{-3} \text{ (metres)}^3$$
$$= 8{\cdot}5 \times 10^3 \text{ (centimetres)}^3$$
$$D = 20{\cdot}4 \text{ centimetres}$$

For the hollow shaft, it is most convenient to obtain first a simpler form of the expression for J in terms of the outer diameter D, thus:

$$J = \frac{\pi}{32}(D^4 - (0{\cdot}9D)^4)$$
$$= \frac{\pi}{32}(D^4(1 - (0{\cdot}9)^4)$$
$$= \frac{\pi}{32}D^4(1 - 0{\cdot}656)$$
$$= 0{\cdot}0338D^4$$

Substituting in the formula as before:

$$\frac{(50 \times 10^3)}{0{\cdot}0338D^4} = \frac{(30 \times 10^6)}{\frac{1}{2}D}$$

observing in passing that the maximum radius for the shaft is still $\frac{1}{2}D$, even though the shaft is now a hollow one. Rearranging, we get:

$$D^3 = \frac{50 \times 10^3 \times \frac{1}{2}}{0{\cdot}0338 \times 30 \times 10^6} = 24{\cdot}6 \times 10^{-3} \times 10^6 \text{ (centimetres)}^3$$
$$D = 29{\cdot}1 \text{ centimetres}$$

It is not surprising to find this answer to be greater than the previous one. We would obviously expect a hollow shaft to be less strong, size for size, than a solid one. But, as a method of comparison of the two shafts, let us work out the ratio of weights. This clearly will be the same as the ratio of cross-sectional areas. Denoting the hollow and solid shafts by H and S, respectively:

$$\frac{W_S}{W_H} = \frac{(\pi/4)D_S^2}{(\pi/4)D_H^2(1 - (0{\cdot}9)^2)}$$
$$= \frac{(20{\cdot}4)^2}{(29{\cdot}1)^2(1 - 0{\cdot}81)}$$
$$= 2{\cdot}59$$

So although the solid shaft is approximately only two-thirds the outside diameter of the hollow shaft, it is more than two-and-a-half times its weight.

You can perceive the analogy between the hollow shaft in torsion and the I-beam in bending. The hollow shaft places the material at the region of highest stress, where it is most needed: it is a design for economy of material. For a shaft of the proportions we chose, the *minimum* stress in the material would be 0·9 times the maximum, whereas the minimum stress in the solid shaft would be zero. The fact that hollow shafts in torsion are very much more rarely seen than I-beams in bending is due to several factors. First, there are few applications where the weight of the shaft is significant: a heavy shaft can be supported just by putting more bearings along it. Secondly, the manufacture of a tube is considerably more complicated than the rolling of a solid shaft, and proportionately more expensive, whereas a steel I-beam can be rolled in a mill almost as easily as a solid rectangular bar.

One notable exception to this general rule is the driving shaft of a car, which transmits a high torque at high speed. It is relatively long, and can only be supported at the ends. A heavy shaft in this situation would tend to whip rather like a skipping rope. Aeroplane shafts are usually bored out hollow for lightness. Apart from the question of pure strength, shafts subjected to heavy duties, particularly turbine shafts and marine propeller shafts, are frequently bored out hollow to facilitate the inspection of the material and the detection of flaws.

CHAPTER ELEVEN

PROPERTIES OF FLUIDS AT REST

How do we define fluids? What makes them different from solids? As with our examinations of other phenomena, the aim is to determine the properties of something with a view to predicting with fair accuracy how it will behave under certain prescribed conditions. For example, a township is to be supplied with water from a reservoir situated on an adjacent hill. It is going to be someone's task to calculate the required pipe sizes in order that the town can be supplied at a suitable rate, so that if everyone happens to be filling a bath at seven o'clock on Friday evening, the supply flowing into the town is adequate to cope with this admittedly rather hypothetical demand.

However, we will leave the discussion of such problems to the next chapter. Our present task is to examine the laws which govern the behaviour of a liquid, not when flowing along a pipe, but when at rest, say in a tank. A light-hearted and enthusiastic plumber may contract to supply cold water to every floor of a fifteen-storey block of flats, from a single storage tank in the roof. Unless he is very sure of his hydraulics, he may find that the pressure at the ground floor is sufficient to burst the pipes.

(1) Definition of a Fluid

As always, this problem of definition seems easy until we begin to think carefully. We cheerfully accept liquids such as water and oil as fluids, and materials such as steel and concrete as solids. But there exists a whole range of substances which are difficult to label clearly. Is thin grease a fluid? If dry sand is a solid, why can it be made to flow along a pipe? Why do some things sink into 'solid' earth over a period of time? One definition of a fluid used to be 'a substance which takes up the shape of its containing vessel'. This was all very well, but it made no stipulation concerning *how long* it was to be allowed to do this. We now know that some substances, such as pitch, *will* take up the shape of a containing vessel if given several months in which to do it. So such a definition is unsatisfactory in that it is not sufficiently *definite*.

The definition now used by mechanical engineers is 'a substance which deforms continuously, when subjected to a shear stress'. We know from our earlier discussions that a material subjected to a single load, such as a concrete column carrying a load at the end, is actually subjected to a shear stress. It therefore follows from the above definition that a fluid is something which will not support its own weight. If you try to pile it up so that it is self-supporting, and it gradually and surely, no matter how slowly, subsides, then it must be a fluid. Some rather surprising classifications emerge from this. For a start, all gases now become fluids to the engineer. Indeed, we know that most of the properties of liquids are also possessed by gases, and the motion of a ship through water has much in common with the motion of an aircraft through the air. Also, palpably obvious 'solids' such as certain plastics, and even glass, possess this property of a fluid.

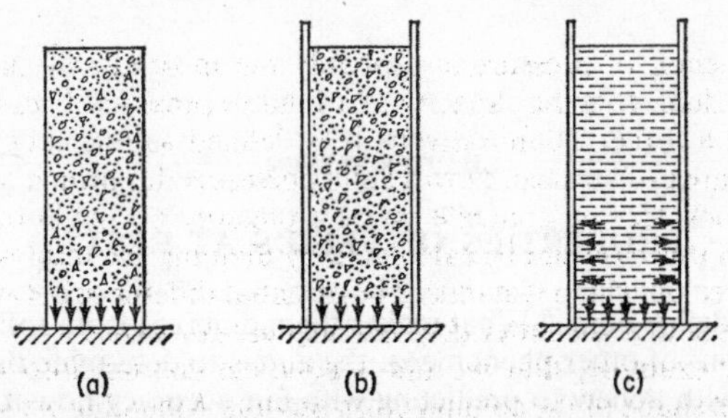

Fig. 115. (*a*) A concrete column exerts a downward pressure only. (*b*) The column is placed inside a steel tank; pressure still acts downwards only. (*c*) The column is replaced in the tank by water; pressure now acts against the walls as well as downwards.

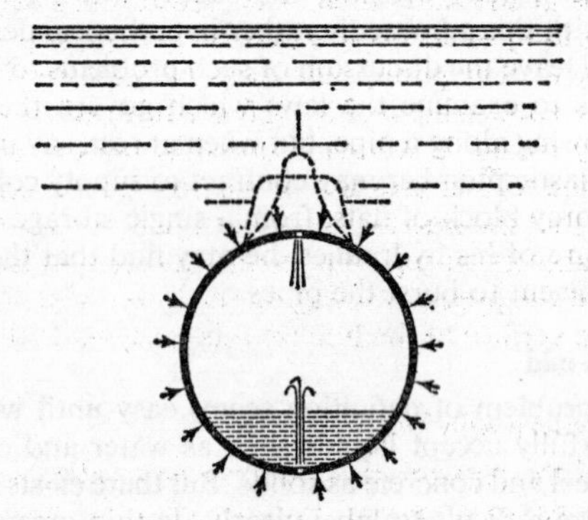

Fig. 116. Pressure in a fluid acts in all directions. Water pours into the submarine through a hole in the top, and also through a hole in the bottom because the pressure is pushing inwards all the way round the hull.

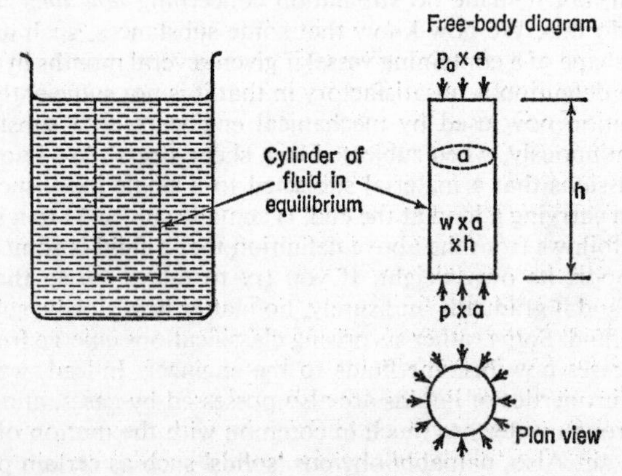

Fig. 117. Forces acting on a cylindrical sample of a fluid.

(2) Pressure

The phenomenon of pressure is probably the most clearly distinguishing hall-mark of a fluid. It is the capacity of a fluid to press itself on to a surface, and thus exert a force. Quantitatively, it is defined as intensity of force, or force per unit area of surface. Now a solid possesses this property in the sense that a solid block of, say, concrete exerts a pressure on the floor beneath it. The pressure in this case can be calculated by dividing the total weight of the block by the area on which it stands. The essential difference between this and a fluid pressure is that the latter *does not only act downwards.*

Imagine, for example, that we put a concrete column into a steel tank having the exact shape of the column cross-section. Assuming we have not had to force it in, the concrete will still only exert a pressure downwards, and no pressure at all against the sides of the tank. But if we replace the concrete column with water, as in Fig. 115 (*c*), this *does* push against the vertical sides of the tank. Moreover, the pressure existing in the water (or any other fluid) acts in *all* directions at the same time. The pressure at a depth of, say, two miles in the sea does not just act downwards, but will press perpendicularly on any surface existing at that point, regardless of the direction in which the surface is lying. The submarine in Fig. 116 is subjected to an inward radial pressure all the way round. If a hole is bored in the top of the submarine, water will rush in due to this pressure. If we now bore a second hole in the bottom in the hope that the water will run out again, we shall be disconcerted to see that water rushes in through the second hole just as fast as through the first, showing us that the pressure acting on the lower face of the vessel is still perpendicular to the surface of the hull (in this case, vertically upwards).

(3) Variation of Pressure with Depth

The magnitude of the pressure in a fluid is easily calculated. Let us take a tank of fluid and consider a vertical cylinder of it, having a cross-sectional area a and height h, as shown in Fig. 117. In the manner outlined in earlier chapters, we can draw a free-body diagram for this cylinder. We now have to indicate carefully all the forces acting on this cylinder, making sure we put them in the right direction.

First, there will be a pressure acting on the upper surface, due to the atmosphere. This acts downwards on the upper surface. Calling the atmospheric pressure p_a, the corresponding down-force is $p_a \times a$.

Secondly, there will be a pressure exerted on the lower surface of the cylinder by the surrounding fluid. This is the pressure we want to find. It will act upwards, i.e. perpendicular to the under-surface of the cylinder. Calling the pressure p, the corresponding force will be $p \times a$.

Thirdly, there will be the weight of the cylinder of fluid; this of course acts downwards. The weight per unit volume of the fluid is called its **specific weight**. Using w as the symbol for specific weight, the total weight of the cylinder will be $w \times a \times h$ (since the volume of the cylinder is $a \times h$).

Fourthly, there will be pressure acting all the way round the curved surface of the cylinder, as shown in the plan view. This pressure will vary, and we do not know much about it yet. However, because it must everywhere act horizontally, the corresponding forces will not affect the equilibrium of *vertical* forces on the cylinder.

We may write our equation of vertical equilibrium in terms of the three other forces thus:

$$p_a \times a + w \times a \times h - p \times a = 0$$

from which, cancelling and rearranging, we obtain:

$$p = p_a + wh$$

Since p_a is the existing pressure at the surface, we have proved that the *increase* of pressure due to the weight of the fluid is given by the product of depth and specific weight.

This formula we have derived is extremely important. It tells us that the amount of fluid does not affect the pressure. Thus the pressure at the bottom of each container shown in Fig. 118 is the same, although the *force* due to this pressure is much greater in the right-hand vessel.

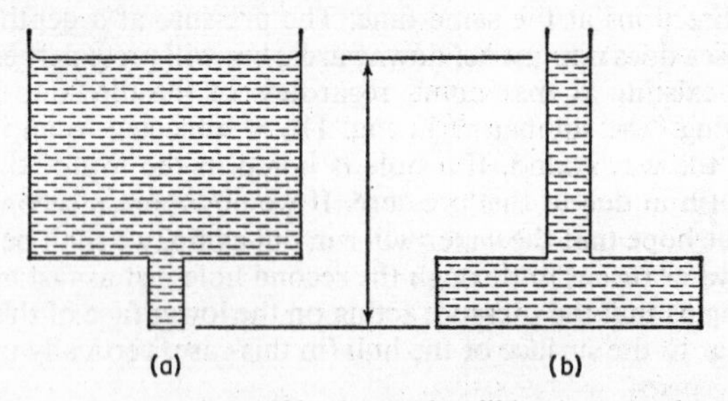

Fig. 118. The pressure at the base of both vessels will be the same, because the depth is the same in both. But the force on the base of (*b*) will be much greater than that on (*a*).

(4) Measurement of Pressure

The formula $p = p_a + wh$ gives us a method for actually measuring pressure. If we have a pipe containing a gas under pressure, and we bore a small hole in the side and then connect to this hole a glass U-tube containing water, the pressure will force the water part of the way round the bend of the tube. The arrangement is shown in Fig. 119. The pressure due to the gas is the same as that due to the difference of the lengths of columns in the two halves of the tube.

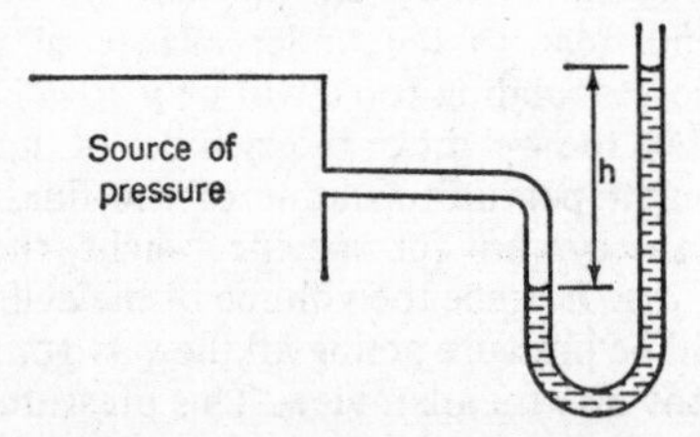

Fig. 119. The manometer registers a difference of pressure by a difference of levels of a liquid in a U-tube. The difference of pressure between inside the tank and outside is $w \times h$, where w is the specific weight of the liquid in the tube.

Such a device is called a **manometer.** We shall see that if water is used in the tube, it is only satisfactory for measuring very small pressures. It is quite common practice to specify the pressure so determined in terms of the manometer reading: for example, a pressure of 20 centimetres of water. Where large pressures are to be found, which would result in inconveniently long columns of water in the tube, a liquid of greater specific weight (usually mercury) may be used instead.

If we could now take our manometer and remove the pressure on the open end due to the atmosphere, the instrument would still indicate pressure, but a much higher one. Instead of indicating the difference between the pressure in the pipe and outside, it would indicate the difference between the pipe pressure and zero. This is known as the **absolute pressure.** One would not normally do this, as the value can be obtained by adding the pressure of the atmosphere (which is known) to the manometer reading, but it gives us a method of actually measuring the atmospheric pressure. If, having removed the atmospheric pressure from the open end of the tube, we now seal it off and connect the other end of the tube to the open air, the manometer will indicate the pressure due to the atmosphere. This modified manometer is shown in Fig. 120 and is called a **barometer.**

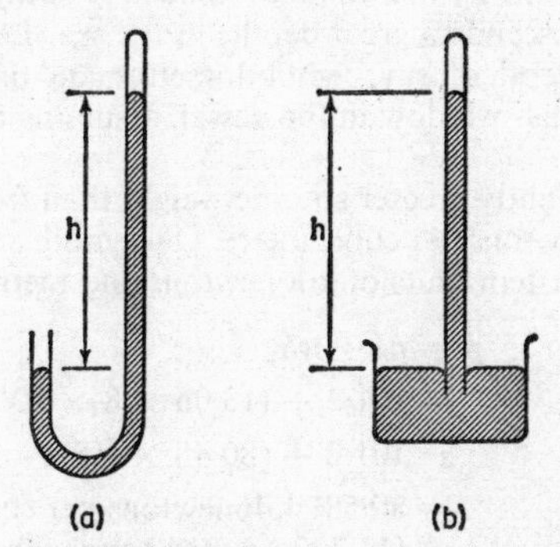

Fig. 120. A barometer is a manometer with one tube sealed and evacuated, and the other open to air. It can be a U-tube as shown at (*a*), but is usually a straight tube dipping in a small vessel, as at (*b*). In both types, the liquid used is mercury, and *h* is the 'height' of the barometer.

(5) The Barometer

The average pressure of the atmosphere is 101·2 kilonewtons per square metre (14·7 pounds per square inch). Let us see what the height of the column of a barometer will be, first if mercury is used as the fluid and, secondly, if water is used.

Mercury has a **specific gravity** of 13·6, which means that it is this many times heavier than water, which has a specific weight of 9·81 kilonewtons per cubic metre (62·4 pounds per cubic foot). Using our formula:

$$p = p_a + wh$$

p_a is here zero because we have zero pressure at the top of our column. p will be 101·2 kN m^{-2}. Substituting these values in the formula, and taking care to work in consistent units of newtons and metres:

$$101{\cdot}2 \times 10^3 = 0 + (9{\cdot}81 \times 10^3 \times 13{\cdot}6) \times h$$

from which

$$h = 0{\cdot}758 \text{ metres (29·9 inches)}$$

On the other hand, using water in the tube:

$$101{\cdot}2 \times 10^3 = 0 + 9{\cdot}81 \times 10^3 \times h$$

giving

$$h = 10{\cdot}32 \text{ metres (34 feet)}$$

These answers should explain why mercury is used in preference to water for filling a barometer. Water barometers have been constructed, but only as scientific curiosities.

(6) Pressure and Force

Let us take two widely differing pressure situations: a bathysphere diving under the ocean, and the air in a ventilation duct. A bathysphere is a spherical vessel designed to descend to great depths in the sea. Let us see what would be the pressure at a depth of, say, eight kilometres, and find the corresponding force exerted on a glass window in the vessel, assuming a circular window of diameter 25 cm.

Sea water has a slightly greater specific weight than fresh water. The value is about 10·06 kilonewtons per cubic metre. Once more applying our formula, and working in consistent units of kilonewtons and metres:

$$\begin{aligned} p &= p_a + wh \\ &= 101{\cdot}2 + (10{\cdot}06 \times 8 \times 10^3) \\ &= 101{\cdot}2 + (80{\cdot}48 \times 10^3) \\ &= 80{\cdot}581 \text{ kilonewtons per square metre} \\ &\quad \text{(11 700 pounds per square inch)} \end{aligned}$$

In a case such as this, we can clearly ignore the effect of the atmospheric pressure at the surface.

The force on the window will be this pressure multiplied by the area of the window. Working now in kilonewtons and metres:

$$\begin{aligned} \text{Force} &= 80{\cdot}581 \times 3{\cdot}14 \times 0{\cdot}125 \times 0{\cdot}125 \\ &= 3960 \text{ kN (397 tons)} \end{aligned}$$

A glass window capable of withstanding this force requires some careful design if a disaster is to be avoided. The windows in Prof. Piccard's famous bathysphere were made several centimetres thick.

As an example of a very low pressure, let us now consider the air in a ventilation duct. If a water manometer were connected to such a duct, it would show a fairly typical value of pressure of about 10 centimetres of water. (This, of course, would be the excess pressure over atmospheric.) Imagine such a duct having a cross-section of about 2 metres by 1·5 metres. We shall first calculate the excess pressure inside the duct. Secondly, we shall determine the total force exerted by the excess pressure on a section of the sidewall of the duct 2 metres by 1·5 metres, this being a reasonable size of steel sheet from which the duct might be made.

At the end of Section 5, we found that the atmospheric pressure of 101·2 kilonewtons per square metre is balanced by a column of water of length 10·32 metres. By simple proportion, the pressure of a column of water of height 10 cm is:

$$p = 101{\cdot}2 \times \frac{0{\cdot}1}{10{\cdot}32} = 0{\cdot}98 \text{ kilonewtons per square metre.}$$

Thus, the internal pressure in the duct is (101·2 + 0·98) = 102·18 kilonewtons per square metre. The force on a sideplate is due only to the excess pressure, since the atmospheric pressure acts equally on both sides of the plate.

$$\begin{aligned} \textit{Force} &= \text{pressure} \times \text{area} \\ &= 0{\cdot}98 \times 2 \times 1{\cdot}5 \\ &= 2{\cdot}94 \text{ kilonewtons (656 pounds)} \end{aligned}$$

Such a high load comes as a surprise, considering the modest value of the excess pressure. It has been observed in an actual duct that such a pressure caused the sides of the duct to 'blow out' several inches at the centre, due to the force, and the steel sheets had to be stiffened to withstand it.

(7) Suction

The idea of suction is familiar to us all, but we need to look at it from the rigorous standpoint of dynamics which we have outlined in Chapter Four. The essence of the first law of motion is that no body will move from its state of rest until acted upon by a resultant force. Furthermore, a force almost always acts by direct contact with the body concerned. A train is pulled by the drawbar of the locomotive. A car is driven along a road by the frictional force between tyres and tarmac. The only forces capable of acting remotely are electromagnetic and electrostatic forces, and gravitational attraction. From this, it follows that if we apply suction to the top of a tube, the other end of which is dipping in a tank of water, we cannot apply a *force* to the water in the tank. All we are doing is removing some of the air in the tube, and hence reducing the downward force exerted by the air on the water surface. Why, then, does the water rise?

By reducing the pressure inside the tube, we have, in effect, the same situation as a manometer. Water will rise up the limb of less pressure. The only difference is that the second limb of the manometer is now replaced by the surface of water in the tank. The water rises because the pressure outside forces it up. Actually, it will rise just so far that the pressure at the base of the column (point X in Fig. 121) is the same as the pressure on the outer surface. If, for example, the outer surface were situated in a chamber completely exhausted of air, the water would not rise in the tube at all, no matter how much suction was applied. If, as is the usual case, the outside is open to the ordinary atmospheric pressure, the inner column will rise in proportion to the difference between this pressure and the reduced pressure in the tube. Since this *difference* cannot exceed 101·2 kilonewtons per square metre, the

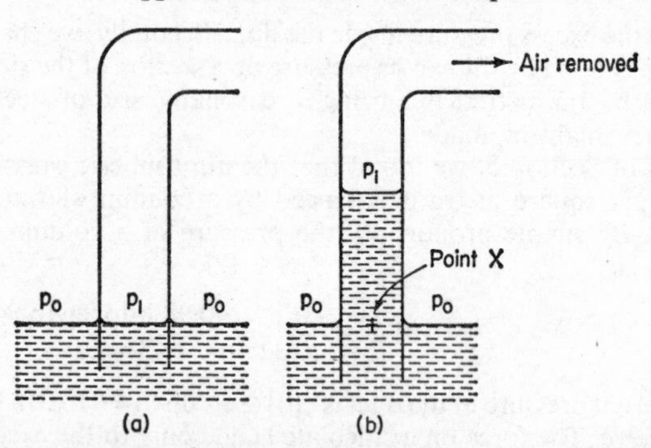

Fig. 121. (*a*) If the pressure p_i in the tube is the same as that outside, the level in the tube will not rise. (*b*) Air has been removed, and p_i is now less than the outside pressure p_o; the water rises in the tube.

water column could never rise higher than 10·32 metres (the figure we calculated in Section 5) no matter how much suction is applied. We would simply have produced a water barometer. It follows that attempts to remove water from a well by means of a suction pump at the top are doomed to failure if the well is more than 10·32 metres deep. The only remedy would be to lower the pump to the bottom, and to force the water up the delivery pipe.

(8) Variable Pressure on a Surface

Our simple calculations so far have assumed that the pressure at a given point in a fluid is constant, and we have calculated force as the product of this constant pressure and the area over which it acts. Think now of pressure acting on the side-wall of a large tank containing water, as in Fig. 122.

If we try to find the force on the wall due to this pressure, we now discover that the pressure is not constant but increases with the depth. Alongside the wall I have shown a graph indicating how the pressure varies with the depth. Ignoring the surface pressure (which acts on both sides of the wall anyway) and using our simple law, we see that the pressure at the water surface will be

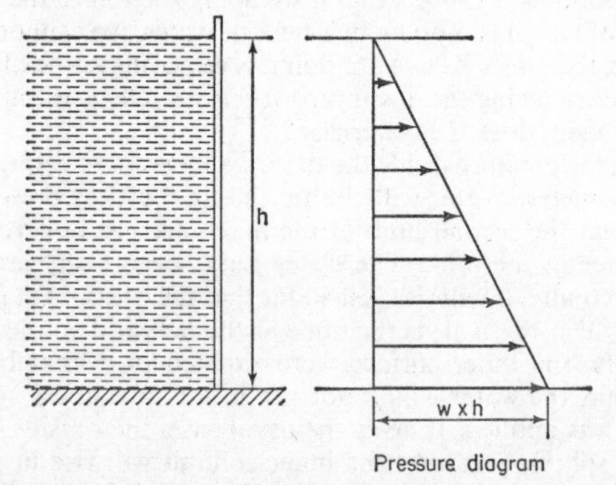

Fig. 122. The pressure of the water against the vertical wall increases in proportion to the depth.

zero. It increases, in proportion to depth, to a maximum value of $w \times h$ (the product of specific weight and depth) at the bottom of the wall.

For a rectangular side-wall, the sensible and correct course would be to find the *average* pressure and assume that the total force is the product of this and the area of the wall. (For the benefit of the more mathematically inclined reader, if the wall is not rectangular the total force can be calculated as the product of the area of the wall and the pressure acting *at the centroid.*) If we were designing a dam, we should not only require to know the total force, but also its line of action. For although the actual force is distributed all over the wall of the dam, its *resultant* (see Chapter Three) has a definite location. We must clearly distinguish here between pressure and force. The average, or mean, pressure is half-way down the dam wall; the resultant force is not. The location of the resultant force is the point where you would have to hold the side-wall against the thrust of the water, assuming there was nothing else to hold it. Observe in Fig. 123 the effect of applying such a force half-way down.

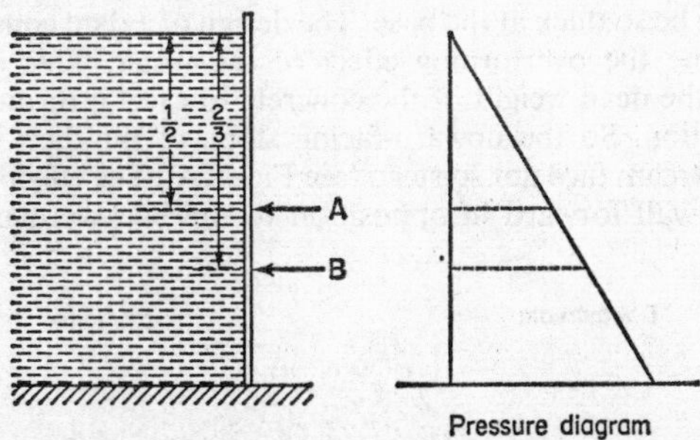

Fig. 123. To hold a vertical wall against a thrust of water, force *A* would not succeed because the lower half is everywhere at a greater pressure than the upper half. Force *B*, at a depth of two-thirds, is correct because the smaller area below carries the greater pressure.

The thrust of the water on the lower half of the plate must greatly exceed that over the upper half: the two half-areas are the same, but the average pressure over the lower half is many times (actually three times) as great as the average pressure over the upper half. Sparing the reader the mathematical proof, it is sufficient to state that for this particular case of a vertical rectangular side-wall extending downwards from the fluid surface, the location of resultant thrust is two-thirds of the depth from the fluid surface.

We can obtain an idea of the magnitude of the forces in such a situation by a simple example. Imagine a large dam 150 metres wide holding water to a depth of 200 metres. We shall determine the pressure at the base of the wall, the total force acting on the wall, and the overturning moment of this force about the base.

The base pressure is determined from our now familiar

$$p = wh$$
$$= 9{\cdot}81 \times 200$$
$$= 1962 \text{ kilonewtons per square metre}$$
(285 pounds per square inch)

The average pressure is one-half this value, i.e. 981 kilonewtons per square metre. The total force will be the product of average pressure and total area of wall, i.e.

$$\begin{aligned} \text{Force} &= 981 \times 200 \times 150 \\ &= 29\,430\,000 \text{ kilonewtons } (2\,960\,000 \text{ tons}) \end{aligned}$$

The turning effect, or moment, will be the product of this force and its distance from the base. Since the resultant acts at a distance from the surface of two-thirds the water depth, the 'moment arm' about the base will be ($\frac{1}{3} \times 200$) metres.

$$\begin{aligned} \text{Moment} &= (\tfrac{1}{3} \times 200) \times 29{\cdot}43 \times 10^6 \\ &= 1960 \times 10^6 \text{ kilonewton metres} \\ &\quad (645 \text{ million ton feet}) \end{aligned}$$

One begins to appreciate why a large dam needs so much concrete, and why it needs to be so thick at the base. The design of a dam consists essentially of counteracting the overturning effect of the water. This is achieved by arranging for the dead weight of the concrete to exert a turning effect in the opposite direction. So the upward-facing slope of the dam is made steep, and the downstream face not so steep (see Fig. 124). The effect is to throw the weight of the wall forward in opposition to the backward moment of the water force.

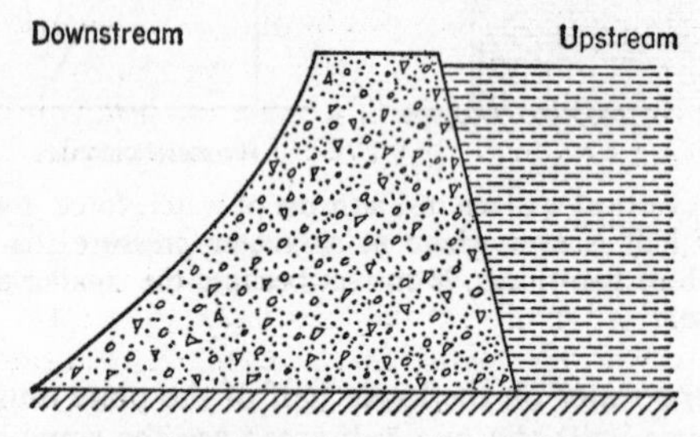

Fig. 124. The centre of gravity of the dam wall is well forward of the centre of the base, so the weight has a turning effect which counteracts the moment of the thrust of the water.

A gravity dam can be considered as an exercise in bending stress. The force due to the water pressure alone exerts a moment on the dam wall, which can thus be thought of as a vertical cantilever; the result of this moment, considered by itself, would be a tensile stress on the upstream face, and a compressive stress on the downstream. Concrete, by itself, is not a suitable material to resist tension; it has to be reinforced with steel. But the weight, being so far forward, exerts a moment of the opposite sense about the base to that of the pressure force. If the dam is correctly designed, this weight-moment will create a compressive stress on the upstream face which is more than sufficient to compensate for the tensile stress due to the pressure.

CHAPTER TWELVE

PROPERTIES OF FLUIDS IN MOTION

The simple calculations of pressure and forces which we have just concluded are not sufficient to solve many other problems associated with fluids. As an example, although we can calculate the pressure and the force on the base of a tank holding water, we have so far no method of calculating the force exerted due to the impact of a jet of water on, say, a turbine blade. This is one typical problem of 'fluid dynamics', as distinct from 'fluid statics' which we examined in the previous chapter. Other typical problems in this field are the measurement of the rate of flow of a liquid or gas in a pipeline, the principle of operation of a carburettor, and, not least, jet propulsion.

In Chapter Five I tried to show that the concept of energy could be extremely valuable in dealing with certain types of problems associated with the motion of solid bodies. The question as to when the energy principle is to be applied instead of Newton's Laws is one which can only be answered by experience. But I propose to show that the application of the principle of conservation of energy is extremely helpful to an understanding of the mechanics of fluids in motion. The problems of the impact forces due to jets impinging on surfaces are best approached using the principle of momentum, which I have already shown to be a modification of Newton's Laws.

(1) Total Energy of a Fluid

When applying the energy method to a solid body, we make use of the principle of conservation of energy. We compute the initial total energy of a body, add to it the result of any work we do *on* the body, subtract from it any work done *by* the body, and assume that what remains is the energy it finally possesses. Only two kinds of energy have come into our examples so far: potential energy due to the elevation of the body above some arbitrary level, and kinetic energy due to any motion possessed by the body; this is because we have not considered instances of bodies which burned to release heat energy, or which blew up to release chemical energy, or which reacted to release atomic energy. But fluids, in even their simplest state, possess another method of storing or absorbing energy, for *pressure is a manifestation of energy*.

Look at the diagram of a tank of water with a pipe and cock shown in Fig. 125. It is clear that here is what we might term an 'energy situation'. If the cock is opened, water will gush out with palpably obvious kinetic energy. I think it is equally clear that it would not do so except for the depth of water (the 'head') in the tank behind it. In terms of energy, the *potential* energy of the fluid due to its height h above the cock is converted to *kinetic* energy when the cock is opened, exactly in the same way as the potential energy of a freely falling body is exchanged for kinetic. When the cock is opened, however, it is *not* the water at the top of the tank which gains kinetic energy, but the water in the pipe which is immediately behind the cock. I have shown three particles of water, which may be imagined as all having the same mass m, at positions A, A' and B. Since B starts to move immediately the cock is opened, it must *already have the energy to do so,* and this is because it is in a state of pressure due to the head of water behind it. The first thing we have to do is to find a method of computing this energy due solely to the pressure of a fluid. We may call this energy 'pressure energy'.

If we take the particle of water of mass m at A and move it to A' we shall change its energy state from potential energy to pressure energy. To perform this operation we need do no work at all upon the particle of fluid, because no force will be required. In fact, under ordinary conditions, particles of water will move about freely throughout the tank without any force being applied.

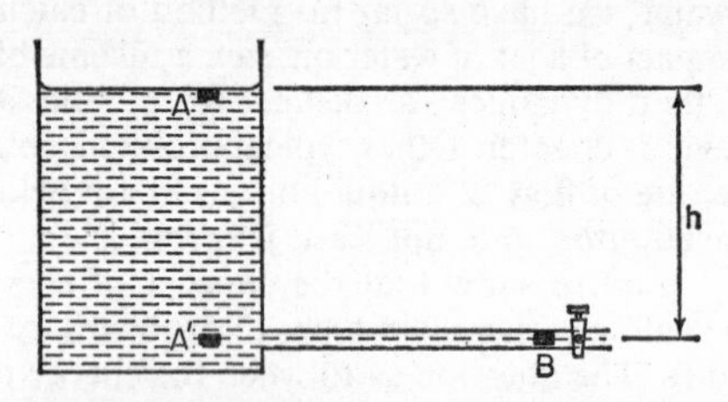

Fig. 125. The particle of fluid at A has potential energy. If the tap is opened, it is the particle of fluid at B that instantly receives kinetic energy. The fluid at B has pressure energy equal to the potential energy at A.

Putting this another way, any given particle of fluid will 'float' in the main body of fluid in the tank, without any tendency to sink or to rise to the surface (discounting any slight thermal effects). So we can with reasonable accuracy apply the principle of conservation of energy and state that the potential energy lost by the particle of fluid at A must be exactly compensated by a gain of pressure energy. Since B is at the same level as A' it is at the same pressure, and thus has the same amount of pressure energy. In Chapter Five we found that a mass m moving through a vertical height h experiences a change of potential energy of magnitude mgh. Thus the gain of pressure energy must have this value.

From Chapter Eleven we know that the pressure p at the point A' can be calculated from the formula:

$$p = wh$$

where p is the *increase* of pressure above that at the surface, h is the head of water, and w is the specific weight of the water, i.e. the weight per unit volume. On page 46 I showed that the mass m and weight W of a body are related by the gravitational acceleration g according to the equation:

$$m = \frac{W}{g}$$

So the specific weight w of the water may be written:

$$w = \rho g$$

where ρ (the Greek letter *rho*) is now the mass per unit volume instead of the weight. (In units, ρ is measured in kilogrammes per cubic metre; w is measured in newtons per cubic metre.) The pressure at A, is thus:

$$p = \rho g h$$

This may be rearranged to give:

$$gh = \frac{p}{\rho}$$

Substituting this in the expression for energy gives:

$$\text{Pressure energy} = mgh = m\frac{p}{\rho}$$

The remaining two quantities—potential and kinetic energy—we have already learned to calculate. We can now write an expression representing the *total* energy of a particle of fluid under any conditions of pressure, velocity and elevation. If we perform no work on this particle, and also permit it to do no work, we can assume that its sum total of energy remains constant. Algebraically:

$$\left(m\frac{p}{\rho} + \tfrac{1}{2}mv^2 + mgh\right) = \text{a constant.}$$

Dividing throughout by mg:

$$\left(\frac{p}{\rho g} + \frac{v^2}{2g} + h\right) = \text{a constant.}$$

mg is the weight of our particle of fluid. Dividing the energy terms by this weight converts the three terms to energy per unit weight of fluid (in specific units, joules per newton) but it may be shown that each of the three terms has the dimension of length. They are called respectively, the **pressure head,** the **velocity head** and the **static head** of the fluid.

This last equation is called **Bernoulli's Equation.** Its principal use is in determining how the pressure in a pipeline varies as the elevation and cross-section of the pipe varies. Let us now consider some of the implications of this very important equation.

(2) Applications of the Energy Equation

Let us assume that water is flowing through the inclined tapered pipe shown in Fig. 126 in the direction A to B. The rate of flow is known to be 185 cubic metres per hour and the pressure at the point A is 70 kilonewtons per square metre (approximately 10·2 pounds per square inch). We shall use the energy equation to calculate the pressure at B. In practice, there is bound to be some loss of energy between A and B, and the hydraulics engineer would almost certainly take it into account, but for the sake of simplicity we shall ignore it. The pressure, velocities and heads at A and B can be denoted by appropriate subscripts. Equating the total energy at A and at B, we can write:

$$\frac{p_A}{\rho g} + \frac{v_A^2}{2g} + h_A = \frac{p_B}{\rho g} + \frac{v_B^2}{2g} + h_B$$

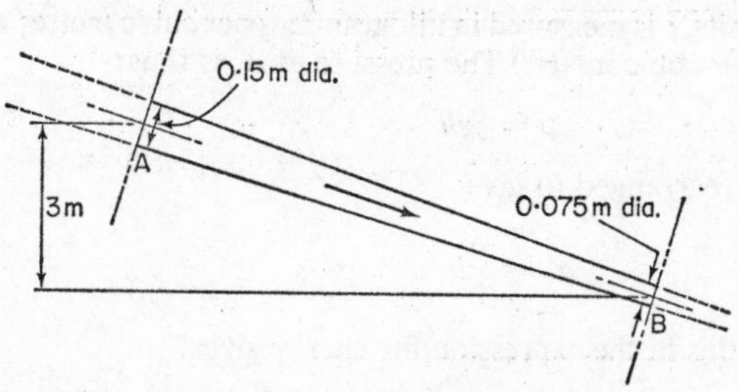

Fig. 126. Tapered pipe down which water is flowing without loss of energy.

We can calculate v_A and v_B because we know the size of pipe at A and B and we know the rate of flow.

$$v_A = \frac{\text{Rate of flow}}{\text{Pipe area at } A}$$
$$= \frac{185}{\frac{1}{4}\pi \times (0{\cdot}15)^2 \times 3600}$$
$$= 2{\cdot}91 \text{ metres per second}$$

and by simple ratio of areas of pipe:

$$v_B = 4v_A = 11{\cdot}64 \text{ metres per second.}$$

The values of h_A and h_B depend upon our choice of datum, but this will not affect the result of the calculation. We therefore choose the most convenient datum for our purposes, which in this case is the one passing through B. Then $h_B = 0$ and $h_A = 3$ metres. ρ, the density of water, is 1000 kilogrammes per cubic metre. Substituting all these values in the equation:

$$\frac{70 \times 10^3}{1000 \times 9{\cdot}81} + \frac{(2{\cdot}91)^2}{2 \times 9{\cdot}81} + 3 = \frac{p_B \times 10^3}{1000 \times 9{\cdot}81} + \frac{(11{\cdot}64)^2}{2 \times 9{\cdot}81} + 0$$

(where p_B in this equation is in kilonewtons per square metre).

Solution of the arithmetic gives a value of p_B of 35·93 kilonewtons per square metre.

Let us now look at the horizontal pipe shown in Fig. 127. This pipe has a

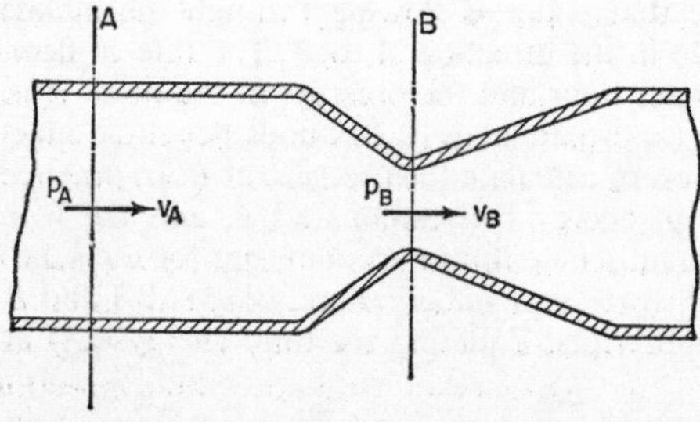

Fig. 127. The constriction in the pipe at B causes an increase of velocity at this point, and a consequent *decrease* of pressure.

section of reduced cross-sectional area at *B*. We shall apply the energy equation to the two points *A* and *B*. The constriction in the pipe at *B* causes the velocity to increase at this point. The situation is similar to the example we have just worked, except that we are now dealing with a horizontal pipe. Because the pipe is horizontal $h_A = h_b$, so we can write the energy equation as:

$$\frac{p_A}{g} + \frac{v_A^2}{2g} = \frac{p_B}{g} + \frac{v_B^2}{2g}$$

(again assuming no loss of energy).

The velocity v_B at the throat must be greater than v_A, and it therefore follows that the pressure p_B at the throat must be *less* than the main pipe pressure p_A. This is in fact true, although it may be rather unexpected; there actually is a *fall* in pressure at a constriction in a horizontal pipe.

A constriction of this type in a pipe is called a **venturi**, and the consequent reduction of pressure is made use of in several devices. One important use is in the measurement of flow in a pipe.

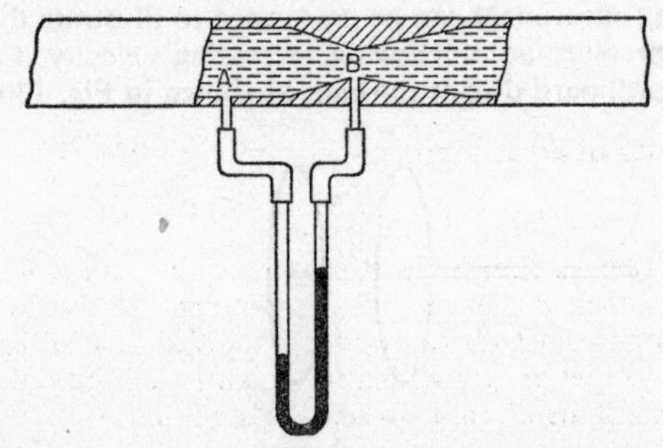

Fig. 128. Pressure difference between the main pipe and the throat causes a change of level of the manometer, which can be used to indicate rate of flow.

If we connect a mercury manometer to the pipe at the two points *A* and *B* in Fig. 128, the mercury in the right-hand tube will be higher than that in the left, owing to the reduced pressure at *B*. The difference of mercury levels will vary according to the velocities at *A* and *B*. The manometer will indicate the pressure *difference* ($p_A - p_B$); if the size of pipe and the size of the throat are known, the velocity and thus the rate of flow can be calculated. But in practice, it is not necessary to carry out any calculations: the manometer scale is *calibrated* for us directly in units of flow rate—much as the stem of a thermometer is graduated in degrees of temperature instead of units of length. The venturi meter is probably one of the commonest methods of measuring the rate of flow in a closed pipe.

A venturi is also used in the carburettor of a petrol engine. The device is shown in Fig. 129. Petrol is maintained at a constant level in the float chamber, which is connected by a pipe to the air duct. Air is sucked through the duct by the induction of the engine's pistons. The connection to the duct is made at the throat of the venturi, and at the same level as the petrol surface in the float chamber (actually, slightly above, to avoid leakage when the

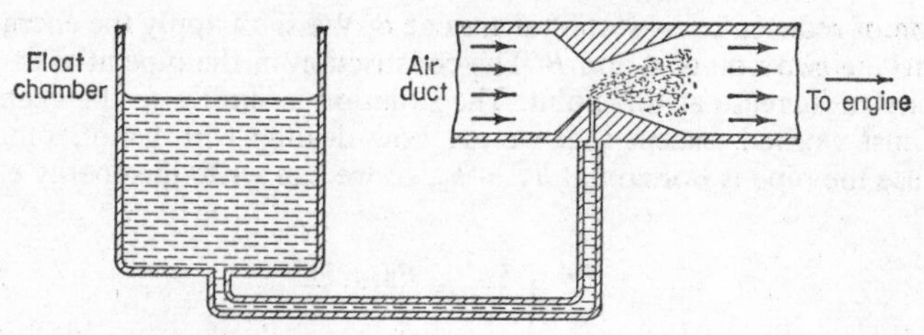

Fig. 129. Petrol is induced into the air duct by the reduction of pressure at the constriction, caused by the flow of air in the duct.

vehicle stands on a slope). When air is sucked in, there is as we have shown a reduction of pressure at the throat. This has two results. First, the petrol is sucked into the air duct. Secondly, the introduction of petrol from the end of a tiny pipe into the rapidly moving air causes the liquid fuel to be 'atomized', or broken up into a mist of tiny droplets, in an ideal state for firing in the cylinder.

An interesting experiment can be performed to illustrate this phenomenon of decreasing pressure accompanying increasing velocity. Connect a small pipe to a flat cardboard disc at the end, as shown in Fig. 130. A milk straw

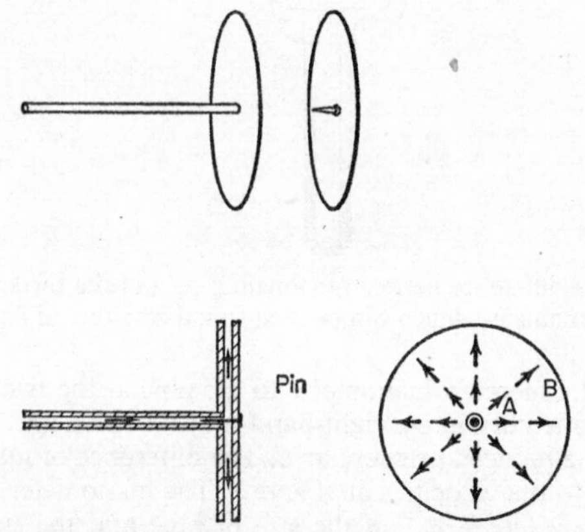

Fig. 130. Blowing through the pipe causes the loose disc to be pulled towards the first disc, not blown away.

will do for the pipe. The pipe must pass right through the disc, so that you can blow through it. Make a second disc of similar size, but with no hole. Blow through the pipe and, at the same time, offer the second disc up to the first. A pin in the middle of the second disc will serve to locate it in the centre, the pin projecting into the open pipe end. You would expect the loose disc to be immediately blown away, and may be surprised to find that (if you blow hard enough) it remains clinging to the first disc. Why is this?

The explanation can be found by examining the flow of air along the pipe and between the discs. You can see that as the air flows outwards between the discs, it must go slower the further it gets out from the centre, because the same quantity of air passing outwards across a circle at, say, *A* crosses a smaller space in the same time than does the same quantity passing a circle at *B*. We have seen that a reduction of speed is accompanied by an increase of pressure. So the pressure must be highest at the outer section of the discs and lowest at the centre. But the pressure at the outer section is ordinary atmospheric pressure. So the unsupported disc has atmospheric pressure acting all over its outer surface, and something less than atmospheric pressure over the inner surface. The result is a net force pushing the disc towards the other one.

A similar demonstration can be performed by connecting the stem of a funnel to a water-tap, turning the tap on so that water flows through the inverted funnel, and then pushing a table-tennis ball into the inverted mouth of the funnel. As long as the water flows, the ball will be sucked into the mouth of the funnel.

Yet another variation is the trick of balancing balls on a vertical jet of water or air. Such a jet tends naturally to spread outwards slightly to the shape of a cone, and this spreading-out causes, in the same manner as the other examples, a reduction of speed and a consequent increase of pressure. The region at the centre of the cone is therefore a region of low pressure as compared with the atmospheric pressure surrounding it, and the ball will be held there, as the table-tennis ball is held in the funnel.

(3) Pipe Friction

Let us now go back to the energy equation we derived in Section 1 for the flow in a pipe. If we now take into account the fact that some of the original energy of the fluid is lost, i.e. the fluid arrives at *B* with less energy than it starts off with at *A*, the equation will have to be rewritten thus:

$$\frac{p_A}{g} + \frac{v_A^2}{2g} + h_A = \frac{p_B}{g} + \frac{v_B^2}{2g} + h_B + \text{Loss}$$

Comparison of this equation with the previous one, using the same data for pressure, flow rate and elevation of pipe, shows that, if there is a known loss, the calculated value of p_B must be less than the value we obtained. In other words, the pressure is reduced by the energy loss in the pipe.

The accurate computation of energy loss in a pipe is very complicated, but simple formulae are available to calculate approximate losses. By far the greatest part of the energy loss is attributable to what is rather loosely called 'pipe friction', and this is dependent partly on the state of the inside of the pipe, partly on the velocity of flow, and partly on the length. The higher the flow rate, the greater the friction loss. This means that if we pass the same flow rate through two pipes of equal length, one having a large bore and the other a small bore, the pressure loss due to friction will be greater in the smaller bore pipe. The implications of this are very numerous, and we can consider only two of them.

Take the case of a township supplied from a reservoir on a nearby hill. The rate of flow is then determined by the size of the town, and the special requirements of its inhabitants, and can be calculated with reasonable accuracy. If we are to rely only on the reservoir head to supply the water, we must provide pipes which are large enough to ensure that the friction loss (for the required pipe velocity) is not greater than the available pressure. Otherwise the rate of flow would be less than the required amount, and the town would have to make do with less, or a booster pump would have to be inserted in the pipeline.

A different kind of problem is that of conveying a liquid, say oil, across a large stretch of country; perhaps several hundred miles. An example is an oil pipeline from a well to a seaport. In such a case, there is not likely to be a 'static head' available, like the reservoir. The oil will have to be pumped all the way. The rate of flow this time is controlled by the output of the well. The choice lies between a pipe of small bore, relatively cheap to install, and a large-bore pipe costing a great deal more. But the pressure loss due to friction in a very long pipe can be tremendous. In this example it must be overcome entirely by the installation of large pumps, which are expensive to install and expensive to run. The decision must be a compromise between these two factors.

An accurate calculation is not possible here, but let us consider a daily delivery of approximately 25 million gallons of oil along 500 miles of a 36 inch-diameter pipe. A rough calculation suggests that the head loss due to friction would be of the order of 5000 feet, and pumps would have to be provided at regular stages down the line to overcome this loss. The work done daily by these pumps would be of the order of 400 to 500 megawatts, involving a probable daily cost of something between £2000 and £4000. Doubling the size of pipe would cut down the cost of pumping (theoretically, at least) to one-sixteenth of this figure, but of course the capital cost of installing such a pipe would be very much higher. In fact, 36 inches is about the largest diameter of pipeline ever used for the transport of oil.

(4) Force of a Jet: Momentum

The problem of calculating the force exerted by a jet of fluid impinging on a fixed surface is one which cannot be solved by energy considerations. Bernoulli's Equation is only applicable to flow in pipes and enclosed ducts, so we have to make use of the principle of momentum. The momentum principle is applied to cases in dynamics where bodies or substances undergo a *sudden* change of velocity. A jet striking a plate is such a case. But the discussion of momentum in Chapter Five applied to solid bodies, and some modification is needed to allow for the non-rigidity of fluids. In the case of a rigid body, the mass is easily defined; with a jet of fluid, we have to consider the change of momentum of a mass flowing in a certain time.

Look at the jet of fluid striking the flat plate shown in Fig. 131. We will call the jet velocity before striking, u. It can be shown experimentally that the jet flows *along* the surface after striking, and does not rebound. There is also only a small loss of energy, so that we may assume the speed of the jet along the plate still to be u. You can see, then, that the momentum of any particle of the fluid (i.e. the product of its mass and its velocity) is unaltered in magnitude, but changed in direction. The effect of the plate is to change the original forwards momentum of the jet, to a sideways momentum. To do this, the plate has to exert on the jet a force in the reverse direction to the original direction of flow. An equal and opposite reaction force is exerted *by* the jet *on* the plate.

If the plate was so shaped as to turn the jet wholly in one direction, it would have to exert a corresponding positive force in that direction. For instance, when water flows round a bend in a large pipe, the pipe bend must be securely anchored to supply the necessary momentum force to change the direction of flow of the water. In the present case, however, where the fluid is distributed radially, equally in all directions, the net momentum in the sideways direction is zero, and no force is exerted sideways.

The momentum principle states that force exerted is equal to the rate of change of momentum. In this case, the force is that of the plate on the jet, which we have seen must act in the opposite direction to the jet flow. The rate

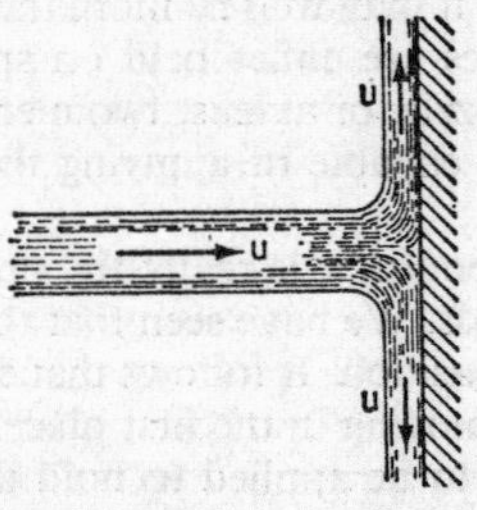

Fig. 131. The velocity of a jet striking a flat plate is unchanged in magnitude, but the direction is turned through 90°.

of change of momentum is the product of the change of forward velocity and the rate of flow (mass per unit time). Writing this in algebraic form:

$$-F = \dot{m}(0 - u)$$

Therefore F is negative because it acts from right to left; $\dot{m}$ is the notation used for the mass flow per unit time. The initial velocity is u, and the final velocity is zero. This gives the simple formula:

$$F = \dot{m}u$$

To obtain an idea of the order of force encountered, let us consider the force of a typical jet from a fire-hose. We can assume a jet diameter of 5 centimetres, and a velocity of perhaps 20 metres per second. This latter figure is u in our formula. The mass flow rate $\dot{m}$ in kilogrammes per second will be the product (area of jet) × (velocity) × (density):

$$\dot{m} = \frac{\pi}{4}\left(\frac{5}{100}\right)^2 \times 20 \times 1000$$

$$= 39{\cdot}3 \text{ kilogrammes per second}$$

taking the standard value of 1000 kilogrammes per cubic metre for the density of water.

Hence:

$$F = \dot{m}u$$

$$= 39{\cdot}3 \times 20$$

$$= 786 \text{ newtons (approx. 176 pounds)}$$

It is not surprising that fire-hoses are sometimes employed effectively for quelling riots. A sudden blow from a powerful jet can easily knock a person over.

The nozzle at the end of the pipe is designed to increase the velocity of the jet to a much greater value than the velocity within the pipe itself (so that the jet will then be capable of reaching great heights and distances). In other words, the *momentum* of the water in the pipe is increased by the nozzle. But in order to do this, the nozzle must apply a *forward* force to the water, and in turn, something or someone must apply a forward force to the nozzle. Although the force will not be quite as great as the impact force we have just calculated (because the water in this case is already moving with some velocity in the required direction), it may well be more than half of it. You may have observed that such nozzles are either held on special stirrups to resist the backward thrust of the nozzle, or at least two men are detailed to hold them, one man alone not being capable of applying the required force for a sustained period.

The principle of jet propulsion is really the problem of the jet striking a flat plate, turned backwards. We have seen that the plate has to exert a force on a jet if it is to stay in position. It follows that exactly the same force must be required to set the jet moving in the first place. (This is the force we have just discussed, which has to be applied to hold the fire-hose nozzle firmly.) If the jet is generated on a freely mounted truck, this force would act between truck and jet—pushing the jet one way, and the truck the other.

For obvious reasons, this is not a practical method of traction for road vehicles. It has been tried in the past with ships, but has never proved economical. However, for aircraft, it has come into its own at last. Here, the jet is provided by the ejection at extremely high velocity of the products of combustion of a fuel. The reasons why jet propulsion has become so important are complex. It is probably sufficient to state that as aircraft speeds increase, so must the speed of a propeller relative to the air. When this propeller speed exceeds the speed of sound, there is a sudden drop of efficiency of the propeller; the torque required to turn it increases suddenly at this point, which corresponds to an aircraft speed of about 0·8 times the speed of sound. As I indicated in Chapter Five, loss of efficiency means expense; for the high speeds of modern aircraft, jet propulsion is more efficient than propeller drive, and therefore cheaper.

The increase of speed of aircraft also has its effect on the size of the wings and tail-plane. The relative motion of the plane through the air may be thought of as having two effects. The shape of wings and tail-plane deflects the flow of air downwards; this produces an upward force, or lift, which counterbalances the weight of the aircraft and keeps it in the air. But at the same time, this flow, and the displacement of the air by the bulk of the plane, exerts a backward force, or drag, which it is the function of the propulsion unit to overcome. Ideally, the wing and tail-plane surfaces of an aircraft should be capable of producing a very high lift for a very low drag, so that it may be held in the air with the least possible engine thrust.

Both lift and drag increase with increase of speed. But the lift never needs to be very much greater than the weight of the plane, so an aircraft at high speed requires only a relatively small wing area to support it. One can see the trend over the years to larger and larger fuselages and relatively smaller wings. At the time of writing, there is a 'wingless' aircraft in course of development. But of course planes still have to start from rest, from the ground. So we have the rather exasperating situation that a large proportion of the wing is only necessary at the low speeds of taking off and landing: at high speeds, a large wing area becomes an actual liability in the form of extra and unnecessary drag. Hence the concept of the 'swing-wing', an aeroplane having a partially retractable wing surface.

EXERCISES

A few exercises are given below, related to various chapters, to assist the reader in the understanding of the topics discussed in the text.

Chapter 2

1. The piston of a car engine at a certain instant has a vertical upward velocity of 5 m/s, while the car has a forward speed along a straight level track of 10 m/s. What is the speed of the piston relative to the earth?
Ans. 11·18 m/s

2. Calculate the required speed of a car wheel of radius 0·36 m for the car to have a speed of 30 m/s. Show that the velocity of a point at the top of the wheel rim, relative to the road, is twice the speed of the car.
Ans. 795·8 rev/min

3. Use the equations on page 14 to calculate the distance travelled in 5 seconds, and the final velocity, of a vehicle which has an initial velocity of 10 m/s and a retardation (i.e. negative acceleration) of 1·2 m/s^2. Find also the distance travelled, and the time taken, before it comes to rest.
Ans. 35 m; 4 m/s; 41·67 m; 8·33 s

4. A car is travelling at a constant speed of 24 m/s around a circular race track of mean radius 200 m. Calculate (*a*) the time to complete one lap, and (*b*) the magnitude of the centripetal acceleration.
Ans. 52·36 s; 2·88 m/s^2

Chapter 3

1. Three forces are shown, acting at a point in space. Either by resolution, or by drawing the Triangle of Forces, show that the forces are in equilibrium. If the 3 kN force is removed, what will be the magnitude and direction of the resultant of the remaining two forces?
Ans. 3 kN opposite to the 3 kN force

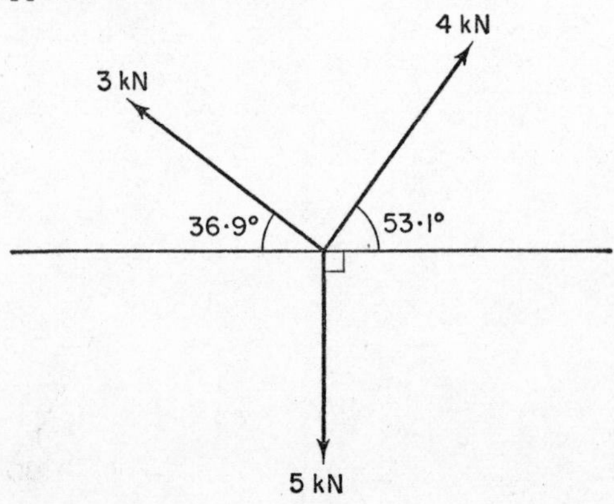

2. A loaded beam is supported at the two points *D* and *G* as shown. Neglecting the weight of the beam itself, determine, by taking moments of forces about *G* and *D* respectively, the magnitudes of the support reactions.

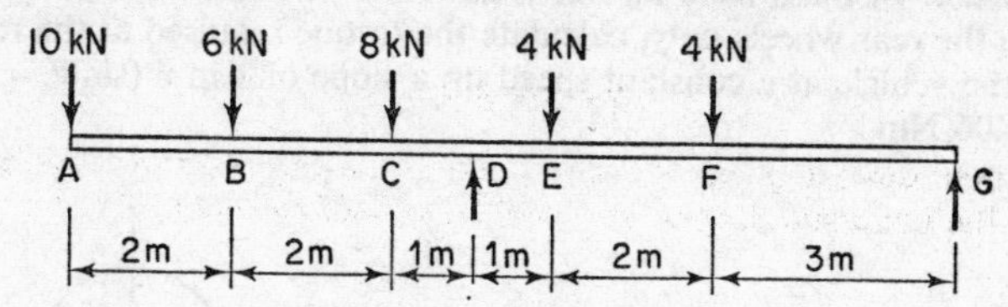

What additional load at *F* would cause the reaction at *G* to be zero? What then would be the reaction at *D*?

Ans. $R_D = 42$ kN; $R_G = -10$ kN (i.e. upwards); 20 kN; 52 kN

3. A steel disc of diameter 200 mm has a circular hole of diameter 25 mm drilled in it, 60 mm off centre, as shown. Locate the position of the centre of gravity of the body.

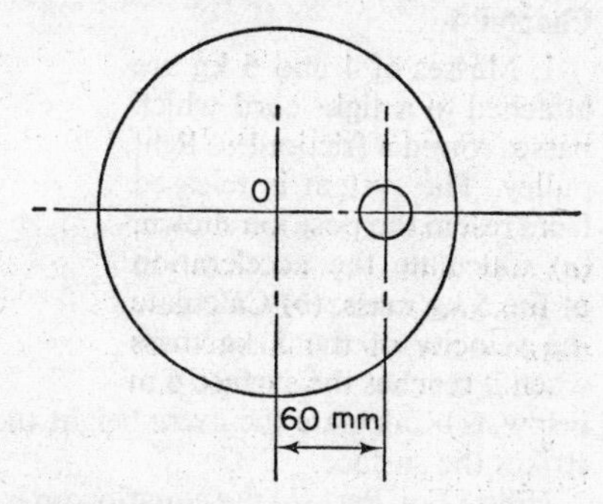

Hint: take 'moments' about O, treating the hole as a negative weight; assume a specific weight *w* and thickness *t*.

Ans. 0·95 mm to left of O

4. A street lamp has a mass of 10 kg. It is to be supported by two wires attached to anchorages 12 m apart in the manner of Fig. 12 on page 28. If the maximum permissible load in the wire is 1000 N, calculate the minimum 'sag' in the suspension.

Ans. 295 mm

5. Determine, both by drawing the polygon of forces, and by resolution, the magnitude of the resultant of the forces shown.

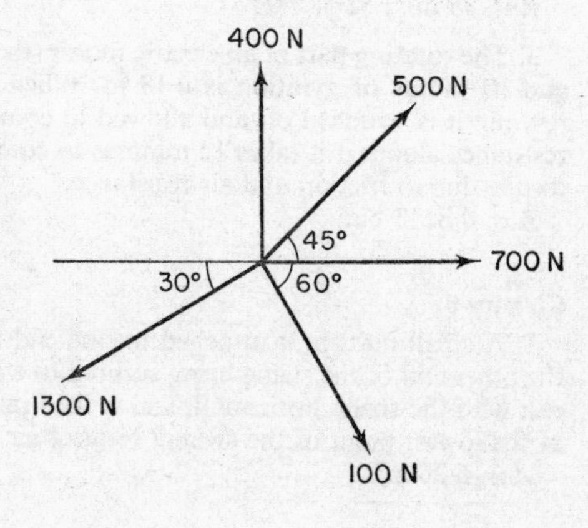

Ans. 28 N. This example illustrates a limitation of the graphical method. The resultant is so small that it cannot be accurately determined graphically.

6. A vehicle of mass 2000 kg has roadwheels of diameter 0·86 m. If it is driven by the rear wheels only, calculate the torque required at the rear axle to drive the vehicle at a constant speed up a slope of 1 in 6 ($\sin \theta = 1/6$)

Ans. 1406 Nm

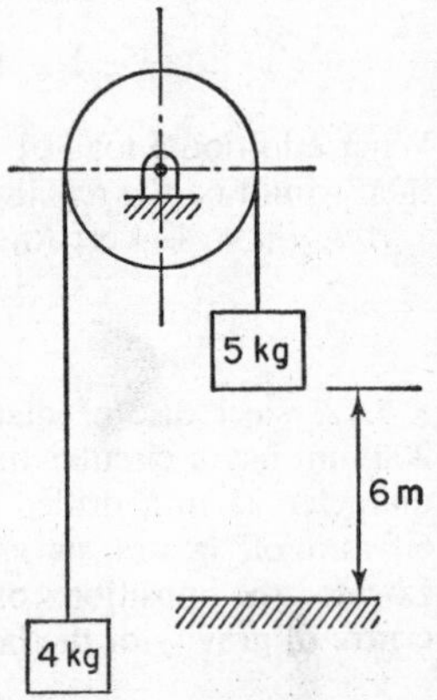

Chapter 4

1. Masses of 4 and 5 kg are attached to a light cord which passes round a frictionless light pulley. The system is released from rest in the position shown. (*a*) Calculate the acceleration of the 5 kg mass. (*b*) Calculate the velocity of the 5 kg mass when it reaches the surface 6 m below. (*c*) Calculate the *extra* height the 4 kg mass rises after the other mass strikes the surface.

5 kg

6 m

4 kg

Hints: For Part (*b*) use equation on page 14. For Part (*c*) the same equations can be used; for this condition, the acceleration of the 4 kg mass is $-g$.

Ans. (*a*) 1·09 m/s²; (*b*) 3·616 m/s; (*c*) 0·667 m

2. A car travels round a circular race track; the track is banked to an angle of 40° and the track mean radius is 150 m. Calculate the speed at which the car must travel around the track if it is not to rely on friction between road and wheels to prevent side-slip. If the coefficient of friction is 0·4, calculate the maximum speed at which the car may travel without side-slip.

Ans. 35 m/s; 52·38 m/s

3. The rotating part of an electric motor (the armature) has a mass of 46 kg and its radius of gyration is 0·18 m. When the motor is running at 1450 rev/min it is switched off and allowed to come to rest under friction and air resistance alone. If it takes 12 minutes to come to rest, calculate the average torque due to friction and air resistance.

Ans. 0·3143 Nm

Chapter 5

1. A small mass m is attached to one end of a light string of length 2 m, the other end of the string being secured to a fixed point. The mass is held at rest with the string horizontal, and is then released. What will be its velocity at the lowest point of the swing? Neglect air resistance.

Ans. 6·26 m/s

2. A car of mass 600 kg is on a slope of length 2000 m and inclination 5° to the horizontal. The engine drives the car with a constant tractive force of 700 N. The resistance to motion due to friction and air may be assumed to be a constant force of 140 N. (*a*) If the vehicle starts from rest at the top of the slope, calculate its velocity at the bottom. (*b*) If the vehicle starts from rest at the bottom of the slope, calculate its velocity at the top.

Hints: the general energy equation is: Initial energy + work done by engine − work done against friction = final energy. For (*a*) initial energy is potential; final energy is kinetic. For (*b*) initial energy = 0; final is potential + kinetic.

Ans. (*a*) 84·57 m/s; (*b*) 17·7 m/s

3. An electrically-driven coal hoist is capable of raising a load of 20 tonnes of coal (1 tonne = 1000 kg) through a height of 300 m at a steady speed of 4 m/s. The hoist is driven through a gear reduction system by eight electric motors, each running at 960 rev/min with a shaft torque of 1·5 kN m. Determine the shaft power of each motor, and the efficiency of the gear-reduction drive system. If each motor takes 172 kW of electric power, what is the efficiency of the motors?

Ans. 150·8 kW; 65%; 87·7%

4. A bullet has a mass of 0·2 kg. It is fired into a stationary sandbag which has a mass of 80 kg, the bullet becoming embedded in the sand. As a result of the impact the sandbag is observed to start moving with a velocity of 3·74 m/s. Calculate the velocity of the bullet.

Ans. 1500 m/s

Chapter 6

1. Given that the coefficient of friction between the 2 kg mass and the inclined plane is 0·4, calculate the value of the second mass, *m*, which is just sufficient to cause the 2 kg mass to move (*a*) up the plane and (*b*) down the plane.

Ans. (*a*) 1·98 kg; (*b*) 0·85 kg

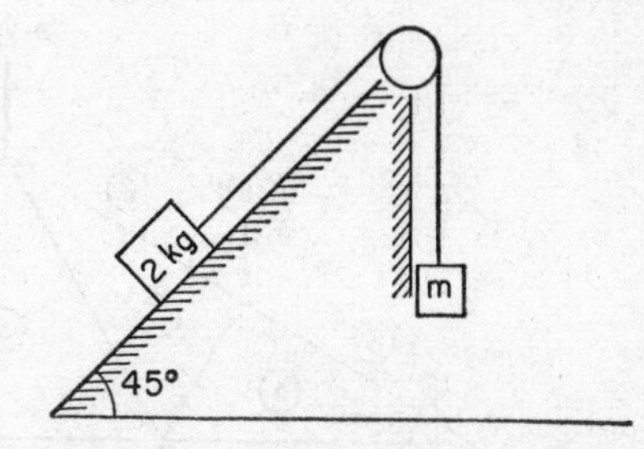

2. The coefficient of friction between mass and plane is 0·2 for both the masses shown. Calculate the least value of mass *m* to cause the 1 kg mass to begin to slide up the plane.

Ans. 1·191 kg

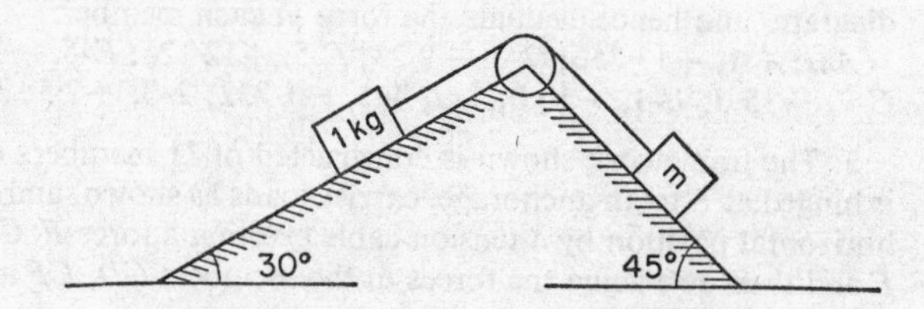

3. A ladder of length 4 m and total weight 200 N leans against a vertical wall. The coefficient of friction between wall and ladder, and also between ground and ladder, is 0·3. Calculate the maximum angle at which the ladder may lean without slipping if it is to support a load of 500 N at the top. (The weight of the ladder itself may be assumed to act at its mid-point.)

Ans. 19·56°

Chapter 7

1. Determine by simple analysis of each joint the forces in the members of the framework shown. All members are either horizontal, or inclined at 60°.

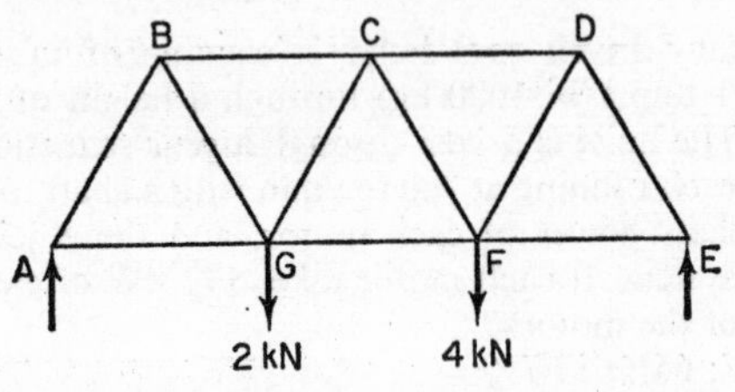

Hints: Calculate reactions at *A* and *E* by taking moments. Then $(AB) \sin 60 = R_A$; $(AB) \cos 60 = (AG)$, etc. At *B* and at *D*, three forces at 120° must be numerically equal.

Ans. AB, BC, −3·079; *CD, DE,* −3·849; *EF,* +1·925; *FG,* +3·464; *AG,* +1·54; *BG,* +3·079; *CG,* −0·77; *CF,* +0·77; *DF,* +3·849 kN

2. The three lower members of the roof truss shown are all the same length; all members are horizontal or inclined at 30° or 60°. Calculate the support

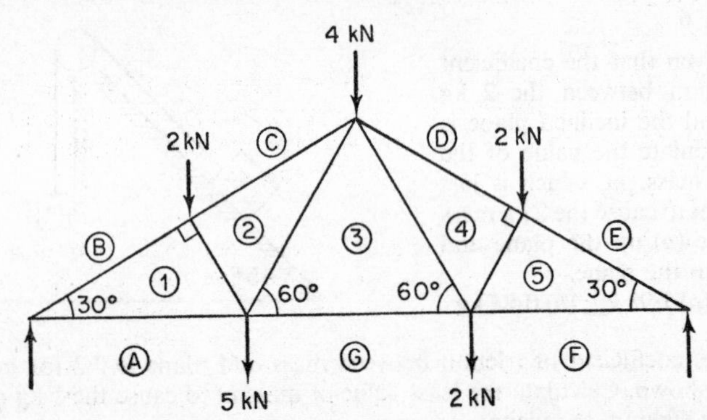

reactions, and, using the notation shown, draw the complete Maxwell diagram, and hence evaluate the force in each member.

Ans. A–1, +13·856; *G*–3, +9·23; *F*–5, +12·124; *E*–5, −14·0; *D*–4, −13·0; *C*–2, −15·0; *B*–1, −16·0; 1–2, 4–5, −1·732; 2–3, +7·5; 3–4, +4·05 kN

3. The framework shown is constructed of 11 members of equal length. It is hinged at *E* to an anchorage, carries loads as shown, and is maintained in a horizontal position by a tension cable exerting a force *F*. Calculate this force *F* and then determine the forces in the members *CD, CF* and *FG*.

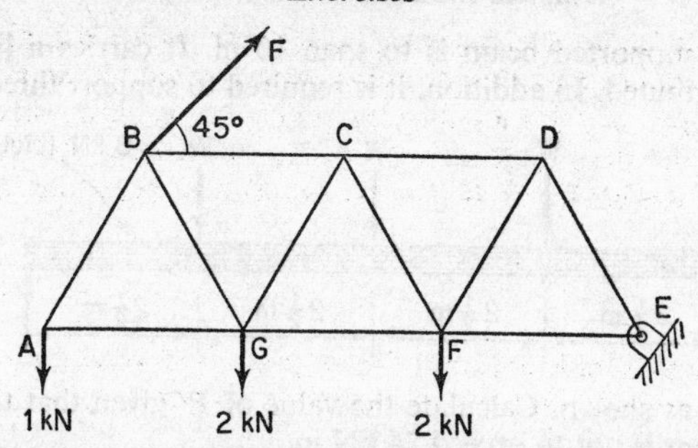

Hints: Calculate F by moments about E. Make a section through the three members stated, and consider equilibrium of the 'cut' portion A–B–C–G. Find FG by moments about C; find CF by equating vertical forces; find CD by equating horizontal forces.

Ans. $F = 3{\cdot}78$ kN; $CD = -2{\cdot}68$ kN; $CF = -0{\cdot}38$ kN; $FG = +0{\cdot}20$ kN

Chapter 8

1. A tie-bar of length 4 m and diameter 25 mm is subjected to an axial pull of 50 kN. Determine the stress and the elongation of the bar. ($E = 207$ GN/m^2)

Ans. 101·9 MN/m^2; 1·968 mm.

2. The following readings were taken during a test on a length of copper wire of diameter 1·75 mm and length 4·5 m.

Tensile Load (N)	0	50	100	150	200
Extension (mm)	0	0·82	1·63	2·47	3·28

Determine the value of E for copper.

Ans. 114 GN/m^2

3. A cylindrical pressure vessel has a mean diameter of 0·6 m and a wall thickness of 4 mm. The material has an ultimate tensile stress of 485 MN/m^2. Determine the maximum permissible internal pressure, adopting a safety factor of 12, based on the value of ultimate stress.

Ans. 538·9 kN/m^2

4. A steel wire of cross-sectional area 4·6 mm^2 is reeled out from a drum and allowed to hang vertically. The density of steel is 7,800 kg/m^3. Calculate the length of wire reeled out if the stress due to the weight of wire itself is not to exceed 100 MN/m^2. Given that E is 205 GN/m^2, what will then be the total stretch of the wire under its own weight?

Hints: recall that weight (N) = mass (kg) × 9.81. The wire will not stretch uniformly; the stretch will be greatest at the top (due to the total weight of wire) and zero at the bottom. Calculate the stretch due to the *average* stress.

Ans. 1306·9 m; 318·8 mm

Chapter 9

1. Calculate the bending moment at the points B, C and D for the loaded beam shown.

Ans. 11, 16, 13 kN m

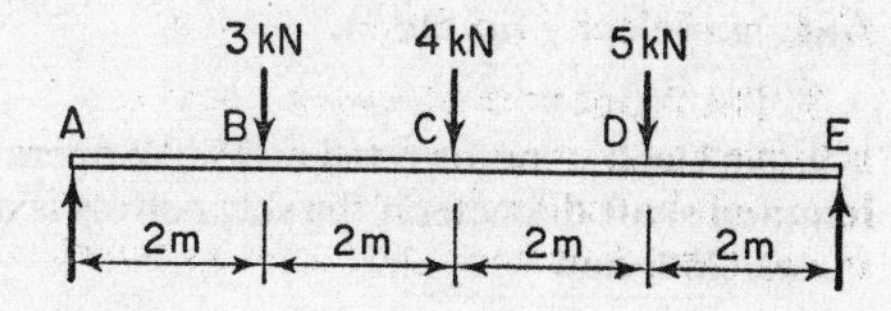

2. A simply-supported beam is to span 10 m. It carries a load of 8 kN uniformly distributed. In addition, it is required to support three equal loads

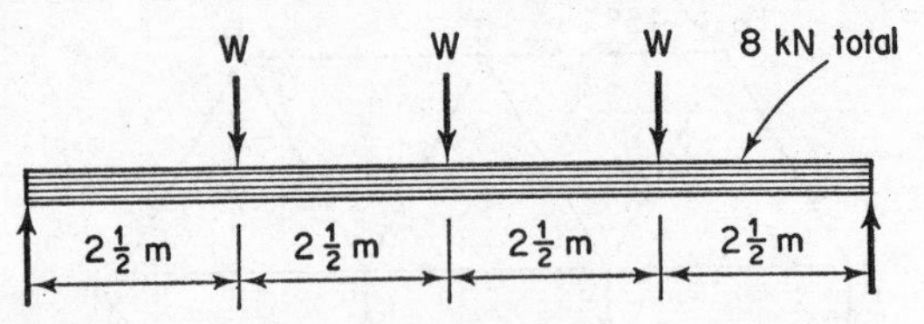

W, positioned as shown. Calculate the value of *W* given that the maximum bending moment is not to exceed 24 kN m.

Hints: Assume the maximum bending moment to be at the centre of the span. Calculate the end reaction, and obtain the bending moment at the span centre in terms of *W* and the distributed load.

Ans. 2·8 kN

3. A timber beam has a rectangular cross-section 250 mm broad by 400 mm deep. It is simply supported over a span of 3 m. Determine the maximum uniformly distributed load it can carry over the whole span, given that the maximum stress is not to exceed 10 MN/m².

Hints: Apply formulae on page 148, using value for *I* on same page.

Ans. 59·26 kN/m

4. A timber beam of rectangular cross-section has a breadth *b* and a depth of 2*b*. It is to be simply supported over a span of 4 m and is required to carry a total uniformly distributed load of 60 kN. Determine the least value of *b* given that the stress due to bending is not to be greater than 12 MN/m².

Hints: Again use formulae on page 148; for maximum bending moment see page 139.

Ans. 155 mm

5. A steel pipe has an outside diameter of 100 mm and a wall thickness of 10 mm. It is used as a simply-supported beam over a span of 2 m. Calculate the maximum possible central concentrated load the beam will support if the stress is not to exceed 120 MN/m². Neglect stress due to the weight of the pipe itself.

Ans. 13·91 kN

Chapter 10

1. Calculate the maximum permissible power which can be transmitted by a solid steel shaft of diameter 40 mm at a speed of 1850 rev/min if the shear stress is not to exceed 50 MN/m². Calculate also the angle of twist of the shaft over a length of 2 m when transmitting this power, given that $G = 80$ GN/m².

Hints: Apply the formula on page 159. Note that Power = Torque × Angular velocity (in rad/s).

Ans. 121 kW; 3·58°

2. An electric motor rated at 25 kW operates at 1440 rev/min. Calculate the required shaft diameter if the shear stress is not to be greater than 45 MN/m².

Ans. 26·6 mm

3. A hollow shaft is to be designed to transmit 375 kW at a speed of 120 rev/min. The shaft bore is to be three-quarters of the outside diameter, and the shear stress in the shaft when transmitting maximum power is not to be greater than 50 MN/m². Calculate the minimum required outside diameter to satisfy this condition, and calculate also the angle of twist of a length of 1 m of the shaft when transmitting the stated power, given that $G = 80$ GN/m².

Hints: Use formula for J on page 159.

Ans. 164 mm; 0·436°

Chapter 11

1. A lock gate has a width of 2·4 m and water is retained on one side to a depth of 3·2 m. Calculate the total force due to the water pressure. Calculate what force would be required when applied to the free vertical edge of the gate, to open it against the force of the water.

Hints: Force = *Average* pressure × Area; average pressure is at half depth.

Ans. 120·5 kN; 60·3 kN

2. A hatch cover in a submarine is circular, with a diameter of 0·6 m. Determine the total force on the cover when it is at a depth of 30 m in sea water which has a specific gravity of 1·02 (i.e. density = 1·02 times that of fresh water). If the door is hinged at a point on the edge and secured by a screw at a point diametrically opposite from the hinge, calculate the force which the screw must exert on the hatch cover in order to resist the force of the water.

Ans. 84·9 kN; 42·5 kN

3. A sluice in a lock gate covers a rectangular hole 0·6 m wide and 0·8 m deep; it is arranged so that the force of the water holds it against the vertical gate. It is raised and lowered by a vertical rod operated by a screw and wheel.

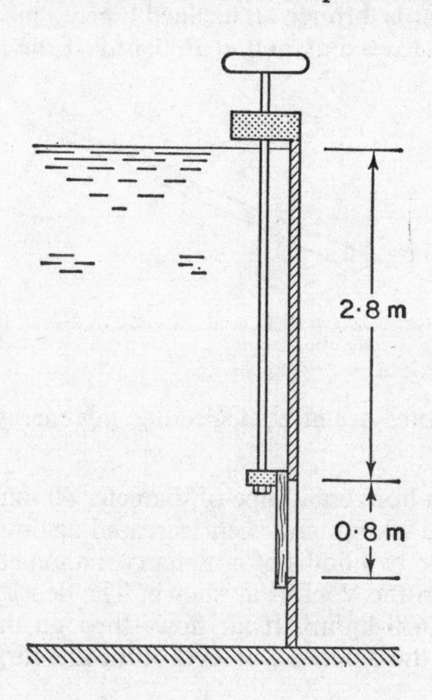

Calculate what force the rod must exert on the gate in order to raise it when the water level is 2·8 m above the top edge of the rectangular hole. The coefficient of friction between the sluice and the lock gate is 0·6.

Hints: Water force = Pressure at centre of sluice × Area; Force in rod = Water force × μ.

Ans. 9,040 N

4. Water is led from a dam through steel pipes of diameter 1·4 m to a power station. At the lowest point, the pipes are 86 m below the level of the dam surface. Calculate the water pressure in the pipe, and thus calculate the necessary pipe thickness, if the ultimate tensile stress for the steel is 450 MN/m² and a factor of safety of 4 is adopted.

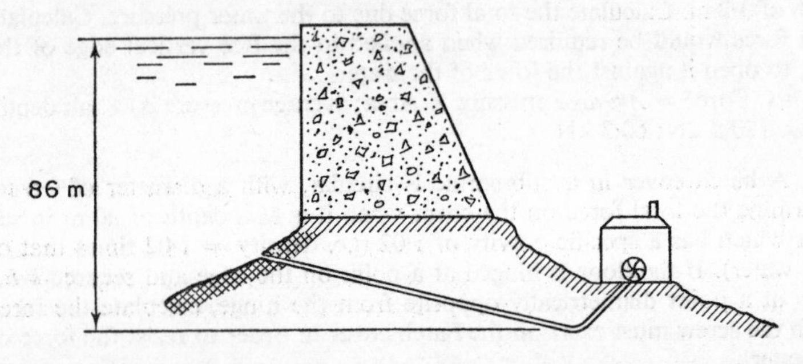

Hints: Use the stress formula on page 128 to calculate the wall thickness of pipe; for safety factor, see page 118.

Ans. 843·7 kN/m²; 5·26 mm

Chapter 12

1. Water flows upwards through an inclined tapered pipe at a rate of 2 m³/s. The pipe diameter at *A* is 0·6 m, and at *B*, 0·4 m. If the pressure at *A* is 300

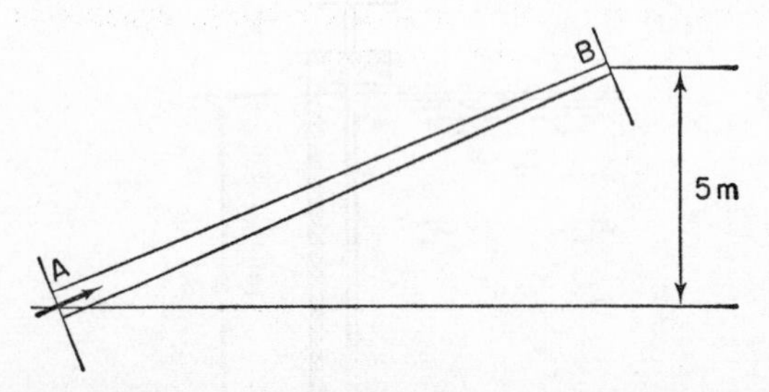

kN/m², calculate the pressure at *B*, neglecting any energy loss from *A* to *B*.

Ans. 149·5 kN/m²

2. Air flows along a horizontal pipe of diameter 40 mm. At one point, the diameter is reduced to 20 mm and then increased again to the nominal size, making a Venturi. The two limbs of a mercury manometer are connected to the points *A* and *B* on the Venturi as shown. The density of the mercury in the manometer is 13,600 kg/m². If air flows through the pipe at a rate of 72 m³/hour, calculate the difference in level of the mercury in the manometer.

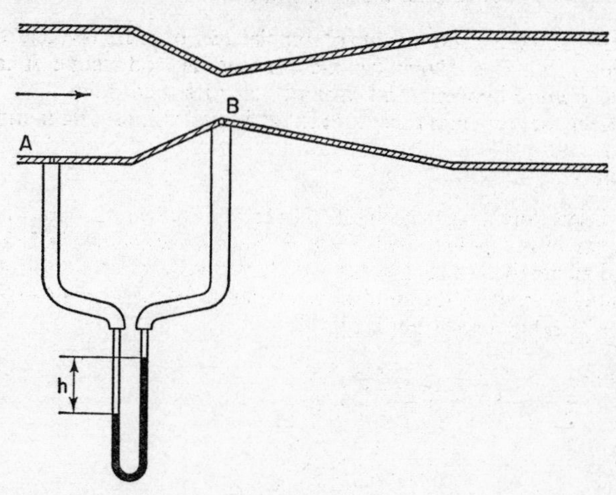

Neglect any energy loss between A and B and assume that the air is incompressible and has a density of 1·23 kg/m^3.

Hints: The two velocities can be calculated (v = rate of flow ÷ area); hence the energy equation can be used to calculate the pressure *drop* $A \rightarrow B$. This pressure drop = ρgh, where ρ is the density of the mercury and h is the height of the column.

Ans. 17·51 mm

3. Oil of specific gravity 0·85 flows along a horizontal pipe of circular cross-section which tapers from 0·6 m diameter at one point to 0·4 m diameter at another. Measurement at the two points indicates a drop of pressure of 65 kN/m^2. Estimate the rate of flow of the oil, neglecting any energy loss.

Hints: Neither of the velocities is known, but velocity at the reduced section (v_2) can be expressed in terms of velocity at the wider section (v_1) because the ratio of the two areas is known. Arrange the energy equation so that one side is *difference* of pressures, which is given.

Ans. 1·735 m^3/s

INDEX

The page references in bold type are of particular importance